DEATH AND ANTI-DEATH
(VOLUME 6)

Death And Anti-Death Series By Ria University Press

DEATH AND ANTI-DEATH, VOLUME 6:
Thirty Years After
Kurt Gödel
(1906-1978)

Charles Tandy, Editor

Ria University Press

www.ria.edu/rup

2008

Printed in the United States of America

Ria University Press   **Palo Alto, California**

Death And Anti-Death, Volume 6:
Thirty Years After Kurt Gödel (1906-1978)

Charles Tandy, Editor

FIRST PUBLISHED 2008

PUBLISHED BY
Ria University Press
PO Box 20170 at Stanford
Palo Alto, California 94309 USA

www.ria.edu/rup

2008 Hardback
ISBN 978-1-934297-03-2

Distributed By Ingram

DEDICATED TO

Rachel Cohon, Ph.D.
Department of Philosophy
University at Albany, SUNY (USA)

Death And Anti-Death Series By Ria University Press

CONTENTS

PREFACE AND ACKNOWLEDGEMENTS

Death And Anti-Death, Volume 6:
Thirty Years After Kurt Gödel (1906-1978)

Volume 6, as indicated by the anthology's subtitle, is in honor of Kurt Gödel (1906-1978). The chapters do not necessarily mention him. The chapters (by professional philosophers and other professional scholars) are directed to issues related to death, life extension, and anti-death, broadly construed. Most of the contributions consist of scholarship unique to this volume. As was the case with all previous volumes in the Death And Anti-Death Series By Ria University Press, the anthology includes an Index as well as an Abstracts section that serves as an extended table of contents. (Volume 6 also includes a section entitled BRIEF COMMUNICATIONS.)

The editor gratefully acknowledges support and assistance from the following:

- I-Fan Chang, Ph.D., President of Fooyin University (Taiwan)

- College of Humanities and Social Science, Fooyin University (Taiwan)

- Research Center for Medical Humanities, Fooyin University (Taiwan)

Death And Anti-Death Series By Ria University Press

BRIEF COMMUNICATIONS

Death And Anti-Death, Volume 6:
Thirty Years After Kurt Gödel (1906-1978)

◆◆

Invitation To Communicate

The Editor invites email to him at <tandy@ria.edu> for possible inclusion in the BRIEF COMMUNICATIONS section of future volumes.

◆◆

Book Review Of Gerard K. O'Neill's
The High Frontier: Human Colonies In Space

Masse Bloomfield

ABSTRACT: In one of THE great books about space colonization, O'Neill offered answers to most of humanity's energy problems and at the same time planned a grand expansion of the human race into the solar system.
KEYWORDS: space colonies; space vehicles; space satellite power stations; space tug; mass driver

[This is a review of the 3rd edition (year 2000) of *The High Frontier: Human Colonies in Space* by the late Gerard K. O'Neill (ISBN 189652267X).] When [the 1st edition of] this book first appeared in 1976, I thought it was one of THE great books about space development. I haven't changed my mind in thirty years.

O'Neill mentions earlier authors on human space habitation. These authors include Edward Everett Hale (1822-1909); Konstantin Tsiolkovsky (1857-1935); in the 1920s, John Bernal, Hermann Oberth, Guido Von Pirquet, and Hermann Noordung; in the 1950s and 1960s, Wernher von Braun, Dandridge Cole and Krafft Ehricke. O'Neill built his space plans on these earlier authors, and he gives them credit in the book.

I was aware of the first edition of *The High Frontier*, but when I went to borrow it from the library, I found there was a third edition. When the first edition was published in 1976, I felt that the program outlined in the book was the way the space program should and would go. I became a fan of O'Neill. O'Neill inspired the creation of the L5 Society which merged with the National Space Institute to form the National Space Society [NSS]. After I read the book in 1976, I thought that NASA was going to follow O'Neill's lead. I was wrong. Where O'Neill described enormous colonies of humans in space, and manufacturing massive solar satellite power stations (SSPS), NASA operates the Space Shuttle and is building the International Space Station. NSS governor and Nobel-prize-winner Freeman Dyson writes in the Introduction that "Believers must accept the fact that O'Neill's reliance on NASA to bring his dream to reality was a grand illusion."

Roger O'Neill in the Preface writes "In 1976, a central message of *The High Frontier* was that the technology already existed to build the space colonies, mass drivers and solar power satellites described in the book. Today, at the beginning of the new millennium, that message is no less true. In several respects technological advances of the last 25 years have further enabled the original vision. However, the human species has collectively not chosen to develop those technologies in a way and to an extent such that we are appreciably closer to the High Frontier today."

Freeman Dyson puts the same thought in a slightly different way. He writes in the Introduction "None of these things (in *The High Frontier*) have happened and none are likely to happen in the next twenty years."

Roger O'Neill goes on to write that "Freeman Dyson ... points out that launch costs for materials from the Earth's surface are probably the single factor most limiting the realization to the vision of *The High Frontier*."

Freeman Dyson also writes "O'Neill's dream makes sense if, and only if, the cost of launching stuff from the ground into space can be drastically reduced."

This book includes the complete first edition without changes. The second part of the book is what I would call interpretations of O'Neill's book. There are six add-on chapters: "Space Robotics" by David P. Gump; "Technology for a Better Future: Space Solar Power an Unlimited Energy Source" by Margo R. Deckard; "Space Solar Power Systems for the 21st Century" by Peter E. Glaser; "Asteroid Resources, Exploitation, and Property and Mineral Rights or Keep Your Laws Off My Asteroid" by John S. Lewis; "A Conspiracy of Dreamers" by Rick N. Tumlinson; and "At Millennium's Eve: The View from 1999" by George Friedman.

It seemed to me that these add-ons did practically nothing to the thrust of getting human settlements in space which was what O'Neill was about. Several were also already outdated with descriptions of cancelled programs.

As I reread the original O'Neill book, it was a booster shot for human colonies in space, the mass driver tug and the SSPS. I felt the book was a draft of a proposal to house billions of people in space and build the SSPS that would correct the Earth's energy and global-warming problem. Energy from space could solve the problems of food production, elimination of pollution, as well as provide unlimited clean energy on Earth. With food production mostly automated, and the SSPS, there would hardly be a limit to the number of humans either on Earth or in space.

O'Neill had the answers to most of humanity's energy problems and at the same time allowed for exponential expansion of the number of human beings in the solar system. He stated the problems confronting mankind when he wrote his book. Those problems have only become worse. He wrote "Within the past decade, four problems have been recognized. All of which relate to the limited size of Earth. They are energy, food, living space and population."

O'Neill's solution in 1976 was "We now have the technological ability to set up large communities in space — communities in which manufacturing, farming, and all other human activities could be carried out."

In a caption under the famous drawing of an O'Neill cylinder it says, "Human colonies in space — not a luxury, but a necessity. Earth is overcrowded, running out of raw materials, in desperate need of a growing energy supply, and being ecologically destroyed. The problems are worse with each passing day, and there are no solutions to be found on Earth itself. Mankind's destiny — its very survival — is in space.... But a commitment is needed, a decision to go for it and the determination to see it through."

Reading O'Neill's book is an inspiration for all of us to foster his plan for humanity. If you haven't read the book, it is still highly worth the time. It is a call for action with a plan for putting humans firmly in space with the option to go to the stars.

[Note:](Includes a CDROM tucked in on the inside back cover which includes videos entitled "The Key" and "The Vision." I did not play the CD so I cannot comment on it.)

http://www.nss.org/resources/books/non_fiction/review_0 08_highfrontier.html

◆◆

An Odd Look At Dying And The Afterlife

Masse Bloomfield

ABSTRACT: In a comparison of functions in living things, dying is a characteristic of both humans and animals. Should animals (having many of the same functions as humans including dying) be going to heaven if humans do?
KEYWORDS: death; dying; afterlife; animal deaths; animal functions

The first aspect of dying can be a comparison of animal functions and human functions. It is difficult for many humans to comprehend the fact that we are animals with animal functions. In addition to animal functions, we do things that no other animals do.

The things we do have in common with animals are:

> Eating
> Sleeping
> Have beating hearts
> Breathing
> Have offspring
> Have sex
> Eliminating
> Bleeding
> Walking and running
> Low level communication (hunger; feelings new born infants express)
> Learning (perhaps only instinctive actions like flight when feeling fear)
> Dying (it seems all higher order animals die)

Then there are activities humans have which none of the other animals have -- or if they do have the activity, it is on a very low level. These activities are:

> Reading
> Writing
> Speaking in sentences
> Use of fingers (perhaps the apes can almost match us but hardly any other animal)
> Using science and technology
> Farming
> Using computers
> Communications (television, radio, newspapers, Internet, telephone)
> Wearing clothes
> Space [extraterrestrial] activities
> Transportation
> Indoor plumbing
> Construction
> Religion

There is a good part of our day taken up in satisfying animal functions and needs. Sleeping, eating and eliminating take up about a third of our normal day. If we exercise, that too is an animal function. The rest of our time is mostly devoted to unique human activities like going to school, working on a job and traveling as well as using our spare time to watch television or to read.

The one function I want to emphasize is one both all higher animals and humans have in common. That function is dying. It is an animal function and a human function. When humans began to use sophisticated language -- that could have been the time when we began to mourn for our loved ones. That date could be from 50,000 years ago to 200,000 years ago.

I know that among most animals, except perhaps the apes, there is some kind of bond between mothers and their offspring which seems to stop when the offspring become able to take of themselves. In humans, the emotional bond between parents and children can last for a lifetime. It is not a universal among humans to kick the child out of the nest when they become adults. There are many things parents do for adult children which no other animal could possibly think about doing. I think it is this attachment between parents and children that has led to the ceremonial burial of dead human beings. The children just do not want their parents thrown away or left to be eaten by predatory animals. We humans are no longer willing to accept the animal practice of leaving the dead as food for predators.

We have developed a large funeral industry to take care of our dead. We even have pauper graves for those people without funds or relatives. Although I must admit some people do provide regular graves for their pet animals.

The major theme I want to propose here, is that we have turned dying, an animal activity, into an activity where we honor the dead. We are unable to think that dead human bodies should become food for dogs and cats. There was a time hundreds of thousands of years ago when that was true.

But we are not animals and we have developed sophisticated religions with morals and rules. The religions have become responsible for overseeing the burial of the human dead.

You may not believe this but no human has ever lived forever. You name a historical figure and death is part of his or her life. Jesus, Buddha, Confucius, Caesar, Joan of Arc, Napoleon, Washington, Lincoln, Roosevelt, all the popes but one, all have died. All of them. Many of the famous people have shrines built to honor their remains. Not one person I know of, either Jesus or an ordinary person, has lived forever. And just as every vertebrate animal, except for the currently living, have also died.

We do have people who believe in such things as reincarnation or ghosts. It is my estimation that those feelings of reincarnation or ghosts, stem directly from our love of parents, friends or heroes. We want to believe that some people are still with us even if they are ghosts.

But I think the more prevalent thinking about the dead is the Christian belief that humans can go to heaven where they have everlasting life. My mind falls into the trap that if humans who are indeed animals, go to heaven, then all vertebrate animals have to go to heaven with us. This would include those animals eaten by predators, or dead animals who happen to die in the ocean and their bodies consumed by sharks or other flesh eating fish.

We are animals and there is no way to change that definition. It is my contention that when we die and when animals die, the very same thing happens. There is a cessation of biological activity. Our systems for respiration, digestion, circulation and reproduction all stop and sensory information and brain function stop. When we die and our cousin animals die, we all are ready for biological or chemical recycling.

Religious faith brings hope and purpose to our lives. It provides a reason to be here whatever God's purpose may be. But this has nothing to do with an afterlife.

If what I have written is true, then there is no heaven, no hell and when humans die, we are no better off than animals. Should we give up those religious beliefs of an afterlife?

◆◆

CONTRIBUTORS TO THIS VOLUME

Death And Anti-Death, Volume 6:
Thirty Years After Kurt Gödel (1906-1978)

Giorgio Baruchello, Ph.D.

Born in Genoa, Italy, Giorgio Baruchello is Associate Professor at the Faculty of Humanities and Social Sciences of the University of Akureyri, Iceland. He studied philosophy in Genoa and Reykjavík, Iceland, and holds a Ph.D. in philosophy from the University of Guelph, Canada. In addition to Akureyri, he has taught at the University of Guelph, at the University of Genoa, and at the International Study Centre of Queen's University at Herstmonceux Castle, UK. His publications encompass several different areas, particularly social philosophy, theory of value and history of philosophy (especially, but not exclusively, early modern French and Italian thought). Since 2005 he edits the Icelandic scholarly e-journal *Nordicum-Mediterraneum*, (http://nome.unak.is).

Daniel A. Dombrowski, Ph.D.

Daniel A. Dombrowski is Professor of Philosophy at Seattle University (USA). He is the author of sixteen books and over a hundred articles in scholarly journals in philosophy, theology, classics, and literature. His latest books are **Rethinking the Ontological Argument: A Neoclassical Theistic Perspective** (New York: Cambridge University Press, 2006) and **Contemporary Athletics and Ancient Greek Ideals** (Chicago: University of Chicago Press, 2009). His main areas of intellectual interest are history of philosophy, philosophy of religion (from a neoclassical or process perspective), and ethics (especially animal rights issues).

Simona Giordano, Ph.D.

Simona Giordano is Senior Lecturer in Bioethics at the University of Manchester (UK). She is the author of **Understanding Eating Disorders**, Oxford University Press, Oxford, 2005. She has published a number of articles on eating disorders, gender identity disorder, obesity and psychopathy.

William Grey, Ph.D.

William Grey is Associate Professor of Philosophy in the Faculty of Arts at the University of Queensland (Australia). He completed an MA at the Australian National University and a PhD at the University of Cambridge. His philosophical interests include metaphysics, bioethics (in particular ethical issues in relation to genetics) and environmental philosophy.

Bill Grote, Ph.D.

Wilfred (Bill) Grote has a PhD in biotechnology from the University of New South Wales (Australia); his MA in ethics is from the University of Queensland (Australia). Bill Grote's research interests include ethical aspects of the application of technological or biochemical interventions into life processes.

John Leslie

John Leslie, University Professor Emeritus of Philosophy at the University of Guelph (Canada), has been Visiting Professor in the Research Department of Philosophy, Australian National University; in the Department of Religious Studies, University of Calgary; and in the Institute of Astrophysics, University of Liège. A Fellow of the Royal Society of Canada, he was its British Academy Exchange Lecturer for 1998. He is known for developing the Platonic theory that the cosmos exists simply because this is ethically required. He is also known for applying the anthropic principle in cosmology, particularly to our place in human population history.

Valerio Lintner

Valerio Lintner is a Reader in European Economics at London Metropolitan University, United Kingdom. He is also a frequent lecturer at a number of European universities, including the University of Montpellier in France and the University of Perugia in Italy. He was previously a Researcher at the European University Institute in Florence. He has published extensively in the area of European Integration.

J. R. Lucas

Fellow of the British Academy. Fellow of Merton College, Oxford, 1960–1996. Author: *The Principles of Politics*, 1966, 1985; *The Concept of Probability*, 1970; *The Freedom of the Will*, 1970; (jointly) *The Nature of Mind*, 1972; (jointly) *The Development of Mind*, 1973; *A Treatise on Time and Space*, 1973; *Essays on Freedom and Grace*, 1976; *Democracy and Participation*, 1976 (Portuguese, 1985); *On Justice*, 1980; *Space, Time and Causality*, 1985; *The Future*, 1989; (jointly) *Spacetime and Electromagnetism*, 1990; *Responsibility*, 1993; (jointly) *Ethical Economics*, 1997; *The Conceptual Roots of Mathematics*, 2000; (jointly) *An Engagement with Plato's Republic*, 2003; *Reason and Reality*, forthcoming in 2009 from Ria University Press. Also see: <http://users.ox.ac.uk/~jrlucas>.

Sir Roger Penrose, Ph.D.

b. Colchester UK 8/8/1931; BSc 1952 (UCL); PhD 1957 (Cambridge); FRS 1972; Knighted 1994; OM 2000. Rouse Ball Professor of Mathematics, Oxford,1973-1998, now Emeritus. Visiting Professor, Penn State University, USA (1993-) and Queen Mary, University of London (2006-). Married (Vanessa 1988, son Max b. 26/5/2000). Prizes/ Medals include Heinemann 1971, Wolf 1988 (with Hawking), DeMorgan 2004, Copley 2008. Scientific papers, books, including *The Emperor's New Mind* (1990 Science Book Prize), *Shadows of the Mind, The Road to Reality*. Research interests: physics/geometry, general relativity/ cosmology (black holes, Big Bang, spinors, conformal techniques), quantum foundations, non-periodic tilings, physical basis of consciousness, twistor theory.

R. Michael Perry, Ph.D.

Dr. Perry (Ph.D., computer science) has worked at Alcor Foundation, a cryonics organization now in Scottsdale, Arizona, since 1987. In spare time he completed a book, **Forever for All**, which deals with scientific and moral issues connected with physical immortality. He is currently working on a revised edition of his book, and on a deeper investigation of the philosophical issues connected with personal identity and survival. Meanwhile, an interest in computers, mathematical programming, and artificial intelligence continues. Dr. Perry is a cofounder and member of the Society for Universal Immortalism which is devoted to one day solving the problem of death in its entirety through a scientific approach.

Charles Tandy, Ph.D.

Dr. Charles Tandy received his Ph.D. in Philosophy of Education from the University of Missouri at Columbia

(USA) before becoming a Visiting Scholar in the Philosophy Department at Stanford University (USA). Dr. Tandy is author or editor of numerous publications, including the *Death And Anti-Death* set of anthologies from Ria University Press. Dr. Tandy, along with Nobel Laureates and others, is a member of the Board of Advisors of the Lifeboat Foundation. Dr. Tandy is Associate Professor of Humanities at Fooyin University (Taiwan) where he serves on the Faculty of History and Philosophy and on the Medical Humanities Research Faculty. Also see: <www.segits.com>.

Clément Vidal

Clément Vidal is a research assistant at the Free University of Brussels (VUB, Brussels, Belgium). His research focuses on the problem of the origin of the universe and its natural laws and constants. He is thus interested in the philosophy of evolutionary cosmology and its implications for our scientific worldview. Moreover he has broad interdisciplinary interests in cognitive sciences, praxeology, complexity sciences, philosophy of science, etc.

ABSTRACTS FOR THIS VOLUME

Death And Anti-Death, Volume 6:
Thirty Years After Kurt Gödel (1906-1978)

Abstract Of Pages 33-52
CHAPTER ONE
Life And Death Economics: A Dialogue
Giorgio Baruchello and Valerio Lintner

Building upon the notions discussed in Giorgio Baruchello's latest contribution to the Death and Anti-Death series, the British economist Valerio Lintner engages with Baruchello in a dialogue about the fundamental assumptions of liberal economics and their implications for the reality of people's lives in the contemporary globalised economic scene. Thus their exchange of views covers key contemporary issues, notably environmental sustainability, and discusses widespread mainstream economic beliefs regarding human motivation, individuality, rationality, wellbeing and happiness.

KEYWORDS: capitalism; globalisation; happiness; liberalisation; monetary policy; motivation; rationality; regulation (of markets); speculation (financial); wellbeing

Abstract Of Pages 53-78
CHAPTER TWO
Charles Hartshorne
Daniel A. Dombrowski

In this encyclopedia-like article, an overview is given of the life and writings of Charles Hartshorne (1897-2000). His method is discussed, along with his crucial distinction between the existence and actuality of God. Further, his theory of value is explicated. A critical evaluation of his view of metaphysics and God is offered. An Afterword presents his views about personal immortality.

KEYWORDS: process philosophy; God; ontological argument; panpsychism; personal immortality

Abstract Of Pages 79-100
CHAPTER THREE
Choosing Death in Cases of Anorexia Nervosa – Should We Ever Let People Die From Anorexia?
PART II

Simona Giordano

Whether or not anorexics should be allowed to die depends not primarily on their competence but on the extent of their suffering and on whether it can be alleviated. If the anorexic has reasonable chance of recovery, competent refusal of treatment can be temporarily overridden. The weak normative role of competence also implies that a patient who has virtually no chance of recovery should be allowed to die in the most peaceful way, even if not all doubts on her competence can be cast aside. This means that nonvoluntary euthanasia for people with anorexia can be an ethical option.

KEYWORDS: anorexia nervosa (anorexics); right to die; role of the family; euthanasia

Abstract Of Pages 101-126
CHAPTER FOUR
The Ethics Of Enhancement

Bill Grote and William Grey

As well as seeking longer lives we also seek better lives. In addition to exploring ethical issues in life extension it is therefore important to explore ethical issues in life enhancement. We examine the nature of enhancement, the therapy-enhancement distinction, and critically examine arguments concerning pharmacological and genetic mind-affecting interventions.

KEYWORDS: life enhancement; psychoactive drugs; genetic enhancement, post-humanism; transhumanism

Abstract Of Pages 127-156
CHAPTER FIVE
Cosmology And Theology
John Leslie

Our universe, as well as having causal orderliness, is fine tuned in life-permitting ways. One and the same parameter, for instance a force strength or a particle mass, often satisfies many different requirements. This could best be explained by the Platonic theory that the cosmos, the sum total of all existence, exists simply because that is ethically required. It could well consist of infinitely many eternal minds, each contemplating the structures of many different universes. Existing among the things contemplated, humans might have immortality. But the human race could soon be extinct, perhaps because of high-energy experiments.

KEYWORDS: cosmogony; universes; anthropic principle; God; Neoplatonism; pantheism; time; doomsday argument

Abstract Of Pages 157-222
CHAPTER SIX
Positive Logicality:
The Development Of Normative Reason
J. R. Lucas

Logic has been misconceived by many philosophers, seeking to make it maximally coercive. They have confined it to analytic deductions which cannot be gainsaid on pain of self-contradiction and general breakdown of communication. Even first-order deductive logic cannot be completely formalised, as Gödel has shown, and there is no hard and fast limit to what should count as logic. We often argue with people who not only share a common language, but a desire to know the truth (in mathematics, about the natural world, or about other people). Such arguments have their own appropriate logic, which is characteristically a two-sided dialogue rather than a monologous exposition of well-formed formulae.

KEYWORDS: logic; deduction; dialogue; holistic; induction; *verstehen*; recursive reasoning; Gödel [Kurt Gödel]; sanction

CHAPTER SEVEN
The Basic Ideas Of Conformal Cyclic Cosmology
Roger Penrose

Conformal Cyclic Cosmology (CCC) provides a new picture, radically different in several respects from currently accepted inflationary cosmology. CCC accounts for the very special nature of the Big Bang (required by the second law of thermodynamics), from a requirement that the universe's conformal (i.e. light-cone) geometry extends smoothly backwards to a pre-Big-Bang "aeon", similar to our own aeon (the currently observed "universe") as part of an endless succession, where the remote, exponentially expanded future of each aeon matches smoothly to a big bang for the next. Black holes play many crucial roles in CCC, both observational and theoretical.

KEYWORDS: cyclic cosmology; conformal geometry; thermodynamics; Weyl curvature; black-hole information paradox; big bang; degrees of freedom; eschatology; quantum geometry; space-time; Tod [Paul Tod]

CHAPTER EIGHT
Deconstructing Deathism:
Personal Immortality As A Desirable Outcome
R. Michael Perry

Mortality is a basic feature of earthly life, yet humans traditionally have been unhappy about it and have sought remedies; the effort continues today, some modern approaches emphasizing the use of science and technology.

Some have tried, on the other hand, to argue that death should be viewed as acceptable and even desirable. Here I rebut such deathist arguments and show how immortality, properly pursued, would be a desirable outcome, a key feature being a growing benevolence and love of others in a community who share in the endless journey.

KEYWORDS: deathism; immortality; resurrection; cryopreservation; benevolence; problem of stagnation; problem of dilution; personal growth; highest happiness

Abstract Of Pages 265-284
CHAPTER NINE
What Mary Knows:
Actual Mentality, Possible Paradigms, Imperative Tasks
Charles Tandy

In part one (of two parts) I show that any purely physical-scientific account of reality must be deficient. Instead, I present a general-ontological framework that should prove fruitful when discussing or resolving philosophic controversies, indeed, I show that the paradigm readily resolves the controversy "Why is there something rather than nothing?" In part two, now informed by the previously established general ontology, I explore the issue of immortality. The analysis leads me to make this claim: Entropy is a fake. Apparently the physical-scientific resurrection of all dead persons is our ethically-required common-task. Suspended-animation, superfast-rocketry, and seg-communities (Self-sufficient Extra-terrestrial Green-habitat communities, or O'Neill communities) are identified as important first steps.

KEYWORDS: all (everything); biostasis (suspended animation); Fedorov [Nikolai Fedorovich Fedorov]; Gödel [Kurt Gödel]; knowledge; mind; O'Neill [Gerard K. O'Neill]; ontology; paradigm; time travel

CHAPTER TEN
The Future Of Scientific Simulations:
From Artificial Life To Artificial Cosmogenesis
Clément Vidal

This philosophical paper tackles the question of the future of simulations in science from a cosmic viewpoint. We argue that it will result in a simulation of an entire universe. The simulation should encompass not only biological evolution, but also physical evolution (a level below) and cultural evolution (a level above). If this simulation could be realized, this would lead to an artificial cosmogenesis. This last direction is argued with a careful speculative philosophical approach, emphasizing the imperative to find a solution to the heat death problem in cosmology. The reader is invited to consult the paper's Annex 1 for an overview of the logical structure of the paper.

KEYWORDS: ALife; cosmological artificial selection; cosmological natural selection; cosmology; fine-tuning; future of science; meduso-anthropic principle; physical eschatology; realization; selfish biocosm hypothesis; simulation

CHAPTER ONE

Life And Death Economics: A Dialogue

Giorgio Baruchello and Valerio Lintner

Lintner: I am starting this exchange of views on the day that Lehman Brothers commercial bank has collapsed, an event which is unprecedented, at least over the last eighty years, and which gives us an indication of just how serious is the crisis of capitalism through which we are passing. Therefore this seems a good time to evaluate just where we stand with regard to economics and the assumptions about human behaviour on which it is based.

It seems to me that the sub-prime mortgage crisis in the USA which has led to today's problems, which incidentally do not just affect bankers and politicians but also the lives and livelihood of most citizens, throws up a number of issues which you explicitly or implicitly cover in your excellent chapter for the fifth volume of the series edited by Charles Tandy.[1] Wouldn't you agree?

Baruchello: I do agree. The collapse of a major commercial bank is no small event. Yet, it is also an event that, quite frankly, could be expected. We might not have known which specific commercial bank was going to go bust, since we do not have access to their accounts , but we could expect that some would fail, and that others might do so fairly soon.

In my previous contribution to the Death and Anti-Death series I discuss at length the dangers of today's deregulated quest for ever-growing profits by the world's financial 'juggernauts' and the ideological myopia underpinning it. This quest resembles very closely the path of action, and of self-destruction, followed by their 'ancestors' in the early 20th century.

After all, the spree of worldwide 'liberalisation' that we have witnessed in the last thirty years or so was meant to do away with institutional 'constrictions' that had been placed upon trade, including the trade in currency and financial commodities. Leading members of the American Republican party, for example, voiced repeatedly and loudly their desire to get rid of all the vestiges of Roosevelt's New Deal, hence of the capacity for State intervention that they required. It is a process begun in the US with Nixon, I would argue, and later followed by many governments worldwide, whether rightwing and leftwing.

Blinded by the neoclassical dogma that wants the freer pursuit of individual profits to translate into collective wellbeing, if not by sheer personal greed, these governments never paused to consider why certain 'constrictions' had been set up in first place. George Soros, Andrew Glyn and yourself did write and speak against this folly; ten years ago, John McMurtry argued that capitalism had already reached its 'cancer stage', for the immune defences of the planetary social body were unable to recognise the biocide invasion and actively cooperated with it by massive doses of further liberalisation. Nevertheless, all these eminent critical voices were ignored, underplayed, or contrasted. And if you stop taking your medication against a certain disease, you are much more likely to catch it again.

What is going to happen, I presume, is that we will rediscover the medication, or else the disease will eat much of the world's economy and, what is implied and never truly spelled out in the mainstream media, the actual lives of many. Millions' employment and livelihood are now at stake because of other people's decisions and errors—the educated, business-savvy elite of some till-then prosperous, glittering country. At least, in the late 1920s, the Russian Bolsheviks had a powerful propaganda machine that kept some people in the West aware of who was responsible for the West's own faults. How difficult it is to do that in the days of Fox News and Berlusconi!

Incidentally, I do not vent the hope that we may learn the lesson once and for all, because humankind seems tragically prone to repeating the mistakes of previous generations. Rather, what worries me is that the medication the world's economy needs, a mix of socialism and effective political leadership, can be administered in various ways. In the 1930s, for example, the State did intervene and, gradually, rescued the world's economy from its irresponsible 'champions', but in most of Europe that recovery meant the affirmation of the fascist model of government and its bellicose forms of public spending. One thing is to cure a wounded limb with an antiseptic solution; another is to amputate the limb—it is a fairly simple comparison of life-value.

Lintner: It seems as if some of the lessons of the 1920s and 1930s have to an extent been learned. As far as macroeconomic management goes, the Fed cut interest rates repeatedly in early 2008, and the Bush administration showered the population with tax rebates in order to stimulate demand (which incidentally brings a wry smile to many economists, given that this is the sort of pure Keynesianism which the neo-cons have so ridiculed in the past...). And you are right, there is going to be a move towards more regulation in the future, and what is more some massive direct intervention by the US government in capital markets. The plan here is for the US government to buy up the bad securities which are at the basis of the so-called `credit crunch' and thereby re-capitalise the banking system, at a cost, it would seem, of some $700billion. The plan, which needless to say is a controversial one, not least among US Conservatives, is to buy the debt at full rather than current market value, thereby effectively subsidising the banks and exposing taxpayers to considerable risk and possible big losses in the future. It is also important to understand the potentially inflationary effects of such a move. What about moral hazard one may ask?

It seems to me this is a wonderful example of the old French saying 'privatisation of profit, nationalisation of risk'…

Nevertheless, there are undeniable similarities between the early-20th-century experience and today's. For example, we live once again in a world dominated by free market capitalism. Since the demise of central planning in the 1980s, the market has established a virtual monopoly (excuse the unintended pun!) across the world as a form of economic organisation. Economic activity, and in particular the financial aspect of free market capitalism, was substantially de-regulated over subsequent years, according to the neo-liberal economic principles that you expose in your book chapter. These principles have come to dominate thinking among academics and practitioners of the 'dismal science'. We have let the proverbial genie out of the bottle, and now it is difficult to see how it can be put back.

However, I do not think that history repeats itself in an identical manner. I believe today's world economy to be different in some fundamental respects from any previous age. And these different, fundamental respects are actually a further source of worry.

First of all, economic activity has over the last couple of decades escaped from the boundaries of the nation state, where it had been essentially rooted since the days of Adam Smith, and moved to the supranational level. This phenomenon commonly referred to as 'globalisation' is of course is due to a number of factors including improved technology, increased and unrestricted trade, the increased importance of Multinational Corporations (MNCs), accumulations of petrodollars which result from the balance of payments surpluses of the oil producing countries and the increased privatisation of welfare and savings. The implication of this is that the world is now much more interdependent: what happens in one place affects everybody. To an extent this has always been the case. In the 1960s and 1970s it was commonly held that "when the USA sneezes Europe catches a cold".

Today, however, the sneeze is much more infectious, as we have seen from the way in which the sub-prime crisis has spread across the Atlantic and beyond.

In addition, nation states could in earlier times quarantine themselves and administer their own form of 'paracetamol' to mitigate the effects. Capitalism is now much more difficult, if not impossible, to regulate. When, in the nineteenth and most of the twentieth century, capitalism was nationally based, it was relatively straight-forward for nation states to develop regulatory frameworks designed to alter the outcomes of the free market to the socially desirable. In principle, people voted for governments that promised them their preferred outcomes, and the governments legislated accordingly: economic democracy in action. Today we live in a different paradigm: supranational economic activity is impossible to regulate and to police since we do not have a world government or any other body which might be capable of regulation free market capitalism. In addition it is clear that the operation of nationally based macroeconomic policies is much more problematic, as a number of local crisis have shown, from the ERM crisis of 1992 onwards.

Baruchello: But aren't we supposed to have international institutions, including non-governmental ones, that monitor and manage financial flows and banking practices? The Basel Committee, the Financial Stability Forum—not to mention the world's central banks, the Bank for International Settlements or the International Monetary Fund—are at work to prevent massive crises like the one we are witnessing today, or are they not?

Lintner: We actually have an international free-for-all on our hands, the objectives of which are certainly not to protect the interests of the vulnerable, or even the majority, and the outcomes of which are probably de-stabilising and certainly unpredictable.

Hence there was very little available to prevent the subprime crisis: the classic accident waiting to happen. One wonders how many more of these there are, and when the next one will come out of the woodwork. The development of the European Union (EU) and other supranational organisations such as the Association of South Eat Asian Nations (ASEAN) can be viewed as an attempt to develop a competent authority capable of modifying the outcomes of the free market on a regional basis. However, such organisations are still a long way from being practically effective regulators of the free market. This is also an issue of democracy: the bankers and market operators that call the shots are not elected by anybody. Democratically elected governments are increasingly unable to control their own economies, which begs the question of why should people continue to vote for people who cannot deliver on the big issues.[2]

Baruchello: I believe their impotence to be due to the leading beliefs and resulting behaviours of the actual individuals involved, rather than to the existing structures—personal, moral factors, rather than structural, organisational ones. The institutions needed for supervision and regulation are there, yet they are not used, or they are used inadequately. Often they even collude with the most blatantly reckless sources of the havoc we are talking about—the failed response of the immune system discussed by McMurtry. Sometimes, major world leaders and the directors of these institutions plea ignorance or insignificance vis-à-vis the gargantuan market forces that they should be withstanding. Still, minor infringements of trade agreements are detected and punished across the planet, suggesting that we or, at least, the WTO, have both the technologies and the expertise to follow countless transactions taking place every month via telephone cables and the world-wide-web.

Lintner: I think it is a bit of both. Of course, the international structures you mention do exist, but their objective is usually to push the free market agenda on the international stage–certainly that has been one criticism of the WTO. It is arguable that they are part of the problem. In addition, the existing institutions are not capable of regulating the type of corporate behaviour and international capital flows that we have experienced. As a minimum requirement, the role and power of existing institutions needs to be re-visited, and arguably we need a 'new financial architecture' more capable of dealing with the issues we are discussing. Naturally, this requires political will and leadership.

Baruchello: As you know, in the book chapter of mine you refer to, I suggest that the neoclassical paradigm has become a *forma mentis* that prevents people from seeing that which is necessary for collective wellbeing and sustainable growth, and thus from behaving in a truly constructive manner.

Within the myopic boundaries of this *forma mentis*, value is understood merely as money capital, not as life-enhancement. Moreover, it is assumed that this pecuniary value ought to be maximised always and anyway, as a sort of Kantian categorical imperative turned 'Rockfelleresque'. We are somehow 'designed' to increase whichever initial capital we are endowed with, says Smith, and, as his disciples continue, we ought to do it, therefore leaping from a factual observation, whether correct or not, to an economic, moral, and political imperative. This logic of maximisation is further justified insofar as it is believed to be bound to guarantee 'optimal allocation' of resources and 'the wealth of nations' itself, as though Adam Smith's 'invisible hand' were out there for sure, which is far from ascertained and ultimately a matter of religious belief—an aspect of Smith's Protestant economics that contemporary economists seem to have forgotten completely.

Thus, a debatable hypothesis about human nature, indeed a token of Smith's religiously inspired philosophical anthropology, becomes the cornerstone of individual and collective agency, indeed the paradigm of human rationality itself.

Lintner: I think you are spot on. The free market system is essentially based on self-interest and short-termism, or if we are to be ungenerous (and why not?) greed – remember the quote from Keynes that you have on your wall at the University?[3] The sub-prime crisis is a classic example of this. Salespersons sold mortgages to people who could not afford them in order to rack up their own earnings. In this they were encouraged by managers who had their own careers and earnings in mind, in turn encouraged by shareholders who want maximum returns now in order to maximise their own wealth, or in order to keep their jobs if they are fund managers. They were all allowed to do this by a regulatory system which is essentially weak and turns a blind eye as long as returns are good and the major actors are happy. And what is more this very system allowed the mortgage companies to then package these essentially unstable and unreliable loans into financial derivatives and products that were then sold around the world and used as collateral for other deals. A pyramid based on a deck of cards, the bottom layer of which was rotten. Of course the collapse of this deck affects us all in one way or other: we all by necessity have a stake in the system through our pensions, our savings and the like. However, some have a much bigger stake than others, for the demise of national regulation has allowed the free market system to move towards an inevitably more unequal society in which wealth is concentrated in the hands of fewer and fewer people.

This of course has many implications, but what interests me particularly is the motivation of the super-rich, many of whom dominate the international economic scene and are the driving force behind many of the developments we are experiencing. Why is it that people who have so much money they would need ten lifetimes to spend it want to accumulate more and more? And indeed why are they prepared to exploit and endanger the interests of us all (and in particular the hundreds of millions who are living on the breadline and for whom economic crisis is not just an increase in mortgage payments, but the difference between life and death) in order to get enough money to last them twenty lifetimes? What motivates them? Why do they do this rather than concentrate on enjoying what they have got? How does this bring them happiness? It seems to me that until we understand more about this we cannot begin to fathom the new international economic realities.

Baruchello: To date, I believe that the most insightful studies on the mentality of the rich are still Thorstein Veblen's, to which, perhaps, I would add some later reflections by Kenneth Galbraith—two economists, hence colleagues of yours, as a matter of fact!

Veblen observed the wealthy elites of the *belle époque*. He concluded that two patterns of behaviour seemed to characterise them, namely 'conspicuous leisure' and 'conspicuous consumption.' Typically, the rich spent more time in idleness or on vacation than the rest of the population and they threw their money around as much as possible and in as many ways as possible. In this perspective, as Galbraith noted seventy years after Veblen, the rich-filled casinos have served the peculiar end of allowing the rich to lose money in public. And why did they want, and still want, to do all this? Because conspicuous leisure and conspicuous consumption are the two main ways in which the rich can show the world that they are rich.

Lintner: Do you think they were suggesting that the pursuit of profit was an end-in-itself from a subjective point of view as well, in addition to being the defining element of capitalism as an economic system?

Baruchello: I believe they thought it could be so for many people. However, Veblen and Galbraith painted a more articulate picture of the wealthy individual's 'human condition under capitalism', if you allow a rather philosophical expression. Being rich meant then, and still means today, the certainty of gaining social status. Thus, what Veblen and Galbraith ultimately argued is that the pursuit of wealth is instrumental to the pursuit of status, which is something that the non-rich strive for too, and may even attain by virtue of, say, political power or cultural recognition. Yet, in their studies, Veblen and Galbraith maintained an unwavering emphasis on wealth. No other instrument seems to be as effective in obtaining and maintaining social status. The rich may be feared, hated, resented, envied or despised, but they are so by the multitude looking at them from the bottom of the pedestal upon which the rich stand.

In this light, political favours and careers, visible statements of cultural distinction, and all the rest that the rich may have bought with their money, aim at securing their status, whether directly or indirectly. Ministries, party leaderships, seats at the House of Lords, foundations, art galleries, villas in Sardinia or the Bahamas are status symbols. Certainly, as Galbraith observed, the typology of status symbols may vary as often and as quickly as the weather, although there tends to be always a rough distinction between the 'tasteless' and 'flashy' symbols of the parvenu, i.e. the new rich, and the 'sophisticated' and 'subtle' ones of 'old money', i.e. the well-established elite. At times, this distinction is unintended. More often, however, it is the result of an ongoing competition between two or more groups within the larger family of a society's wealthiest members.

Should I go for a big, polluting, uneconomical monster-limousine or a smaller, green, fuel-efficient hybrid car? By opting for either path of conspicuous consumption you side with a certain 'party' within the elite, while at the same time showing that you are well above the *hoi polloi*.

Lintner: It is true that riches are a tool for social advancement and social recognition but, as an economist, I cannot avoid seeing how social advancement and recognition, and the status symbols you are speaking of, are often also a tool for gaining or retaining riches. They can be publicity stunts, long-term investments, or ways to humiliate and destroy one's competitors. For example, recently there have been a spate of rich individuals using their enormous wealth to buy football clubs here in the UK. An example would be Abramovich at Chelsea, but I believe there is also an Icelandic example at West Ham. It doesn't matter how much money you've got, nobody might have heard of you, but if you own a football club you are on the back pages all the time and you become a public figure.[4] Also, wealth seems to me to be much more stable in time than the various status symbols it can buy. If you like, it is the best status symbol there can be. Nothing beats a hefty bank account—as long as people know that you have it, of course. Yet, what interests me, is that there must be some deeper psychological drive at work here, which explains why money, *qua* 'king' status symbol, and its 'vassals', things like yachts and private jets, are accumulated by people who have already enough of them.

Baruchello: Status symbols are a matter of fashion, a dimension of social existence that is always characterised by a tension, a contradiction, between conformity and distinction.
As the sociologist Georg Simmel argued, an individual follows fashion to fit in a group of people, yet s/he wants also to stand out, hence s/he introduces an element of variation in the existing fashion, which may become eventually a new trend altogether to which people conform and modify for the sake of distinction, and so on.

If Simmel is correct, and I believe he is correct, this means also that even the wealthiest few are never entirely pleased with their immense fortunes and what they can buy with it, if there are other very rich persons that can do the same, or even outdo them, which sooner or later is likely to happen.

There may be exceptions, of course, but even if only a minority of the world's most affluent individuals plays this game, then they can affect the lives of millions who, either directly or indirectly, depend on the factories, enterprises, capitals and speculations manoeuvred by this minority.

Lintner: I doubt that we are talking of a minority: look at the size of the worldwide trade in luxury goods, works of art, and top-level real estate, or at the glossy magazines targeted at the super-rich and 'wonna-be-rich'... As for the majority of the population, the irony of course is that the pursuit of individualism seems not bring happiness. Some economists, for example Andrew Oswald at Warwick University, have begun to take an interest in this, and surveys on both sides of the Atlantic suggest that people are no happier now than they were fifty years ago, despite huge increases in National Income. Interestingly enough, reported happiness is highest among the highly educated, women, the young and old (not middle aged people who are directly involved in the 'rat race'), people who are married and retired, those staying at home and those who are self-employed. Material prosperity clearly comes at a cost. It is certainly true that 'it is better to be rich and unhappy than poor and unhappy' (my grandmother), but the 'those who say money can't buy happiness don't know where to shop' (anonymous) brigade don't quite get the whole picture, it seems.

Baruchello: Indeed. Ironic is also the fact that the logic of the struggle with one's peers for status is the same amongst the super-rich as it is amongst the kids in a Brazilian favela or a Nigerian bidonville. As the poor kid steal to have the fanciest Nike trainers in his group, so does the rich banker crave for more, and sometimes steal, in order to display the supreme luxury item of the moment that the others can't buy, whether it is a Picasso, a football team, a younger trophy-wife or trophy-husband, or a mega-yacht. This logic is reproduced at all the levels of the social hierarchy: employers and employees, CEOs and part-time cleaners, aristocrats and plebeians. Yet, instead of wearing an overpriced, mass-produced, coloured plastic wristwatch, the very rich are to boost an outrageously expensive, custom-made, gold-and-diamonds watch.

Besides, in this circus, the very rich set the tone of the whole show. They are the role-models for everybody else in the capitalist society, since no alternative economic order is either praised or permitted (see what has happened to the communist bloc or to Islamic economies). And to make sure that the tone is heard and followed, this form of gluttony is fuelled and refuelled ceaselessly by scientifically-crafted advertising. It is even theorised and justified by neoclassical economists as a natural and good disposition of the human being, just like Adam Smith's presumption concerning the human being's natural and good tendency to augmenting the initial capital available. They dub it the 'non-satiety principle' and that is why I find the word 'gluttony' very appropriate.

Lintner: This is all very interesting and important, and it may explain the motivations of some of the people we are discussing, although I still find it unfathomable how intelligent people can have so little self-analysis and self-awareness.

I guess these kind of attitudes are ingrained in our societies, and have been increasingly so at least since the demise of the post-World War II settlement, which in Europe at least had emphasised the 'social market' and a degree of collective consciousness and responsibilities. One of Thatcher's most significant sayings was 'there is no such thing as society', for example. The attitude right now seems to be that the market has seen off the competition, and cannot be challenged.

Former Labour government minister and now European Commissioner (for trade!) Peter Mandelson, Blair's adviser and confidant, is quoted as saying that he was 'immensely relaxed' at the prospect of the emergence a class of super-rich people in the UK–if you can't beat then, join them! These people clearly set the societal moral and philosophical agenda, which since the 1980s has been firmly based on self interest.

In terms of the analysis of individual behaviour, historically economists have attempted to analyse behaviour and social phenomena, but have not really managed to escape the straightjacket of an economic rationality based on self-interest. The Nobel prize-winning economist Garry Becker is a good example of this. More recently, some economists have toyed with the idea of analysing happiness, but the numbers that are interested in this is very limited, and their work is firmly outside of the mainstream.

Having said that, it is one thing understanding why people behave as they do, and quite another to tolerate the essentially anti-social implications of this kind of behaviour, and so we return to the issue of regulation...

Baruchello: Regulation and culture or, if you like, moral education. The mind is the place where we turn modifiable human arrangements into dogmas, cages and straightjackets; but it is also the place where we can be freed from them.

You mention the analysis of happiness. Much of the horror and folly that we have been eviscerating is due to a largely mistaken notion of happiness, which characterises modernity. This is particularly blatant to a person like me, trained in ancient and medieval philosophy. Back then, as savage and 'unscientific' those times may have been, the mainstream notion of wisdom was tied to the notion of reducing needs, not satisfying wants. 'Non-satiety' was a nightmarish option, which only the child, the hedonist and, in essence, the unwise would choose.

Lintner: Yes, in the UK and some parts of northern Europe we have the puritan tradition, which incidentally is still strongly present in some of the green and alternative schools of thought here. I do have a problem with aspects of this approach; however, as it seems to me that some people here seem to revel in tokenism, depriving themselves and others seemingly for its own sake.

Baruchello: Certainly, these ideas were the offspring of ages in which extreme misery was much more common than today and consolation could be found in sharing poverty rather than in generating more wealth for all. Mass cynicism, early Christianity and many medieval 'heresies' exemplify it. Still, I believe they teach us something of fundamental importance. Material affluence may be important, but it is neither the only nor the most important dimension of human existence, individual and collective. Aristotle and Epicurus, for instance, were regarded as rather relaxed vis-à-vis material goods: they were not ascetic enough for many of their colleagues. Yet, they lived and preached nevertheless a mantra of moderation, reduction of needs and focus on what truly matters in life: peace, health, friendship and the cultivation of spiritual abilities.

An esteemed colleague of mine, German historian Markus Meckl, claims that capitalist societies are characterised in late modernity by a depressing lack of higher ideals, which we face most brutally when we want to tell our children how they should lead their lives. Be successful? Make a lot of money? Become a professional footballer or a TV starlet? Our forefathers had a much more interesting set of things to say: be virtuous and save your soul, serve your God and your country, be a good example. Today, we have all these beautiful material goodies and yet even the super-rich flock into rehab clinics or get caught with crack and heroin in their purse while entering the American Embassy in London. They reflect on the grand scale the far-too-common condition of meaninglessness that pervades modern societies.

The loss of religious belief, but perhaps the unseen faith in Smith's 'invisible hand', has been a sign of emancipation from superstition and oppression, but it came at a cost. 'Rationalisation', as Weber called the modern liberation from ignorance and superstition, brought the whole universe within reach of the calculating human intellect—scientific and economic—thus depriving it of mystery, beauty, and of the awe-inspiring 'otherness' that was commonplace in previous ages. 'Disenchantment', he dubbed it; to the point that we have been trying to re-enchant it with things like scientology and Star Trek's unknown alien species to be discovered!

But there is another aspect that I find most troubling and that connects with the issue of economic rationality that you have just mentioned. What sort of rationality can this be, I wonder, that is leading the world to the brink of ecological collapse? How shallow is this reason that treats the damages operated upon the very basic environmental structures that sustain life as 'externalities', as though those structures were not in fact the most 'internal' dimension imaginable?

Without those structures, life would not be possible; and without life, your clever rational agents would never be able to trade freely whichever goods may lead them to be mutually satisfied and bring about optimal allocation. It is a rationality that seems to favour the short-term gratification of whichever immature yen one may have and be willing to pay for, rather than the long-term satisfaction of well-established needs of human communities across generations, whether anyone may be selling or buying anything in the process. It is a rationality that seems unable to see and deal with life and its essential requirements; and whenever it stumbles into it, it sacrifices it to balance sheets and higher interest rates. I must confess that this rationality looks rather like a grand-scale Freudian 'rationalisation' of base instincts.

As for the super-rich, they operate as role-models of life-destructive consumption. Their private jets, big cars, expensive furs, rare-woods furniture, blood-covered diamonds and many and often empty villas are the ideal horizon towards which the non-rich direct their gaze and, as far as possible, their lifestyle.

Lintner: I share yours views regarding economic rationality, and I am also painfully aware of the number of people out there who 'know the price of everything and the value of nothing'. I was also wondering when we would arrive at the issue of the environment and its compatibility with capitalism. This is clearly the issue of the day since, as you mention, it is pointless to argue the toss about philosophy or economics if there is no planet! Now it seems to me that the problem here is that most economists have been either in denial about climate change, or alternatively have no real answers to the problem. Denial has been increasingly difficult recently, in the face of pretty overwhelming scientific evidence, although enough economists still cling on to the opinion that 'it isn't happening' (the Bush administration), or, if it is, then the market and technology will automatically solve the problem, so why worry. The emphasis of the non-deniers has therefore turned, sometimes reluctantly it would seem, to possible solutions. The suggested way forward has in one way or other involved the market and the price mechanism, in the tradition of the theory of state intervention in the market in the presence of the 'market failure' and the 'externalities' to which you have referred. The approach has been to advocate price increases to reduce the demand for car use, air travel, and other activities which are likely to have a negative effect on the environment (although some other activities like road haulage and even military activity seem mysteriously to have been ignored – one wonders what the carbon footprint of the wars in Iraq and Afghanistan might be…). A variant of this has been the introduction of carbon trading, which in principle attempts to reduce carbon emissions by pricing them – economic agents are given permission to emit carbon, and these permits can then be traded. The problems with carbon trading are both practical and philosophical: firstly, the permits so far assigned have been far too liberal to have any effect on the environment. More centrally, this is an attempt to solve problems by using the very mechanisms which, as we have discussed, have been partly responsible for creating many of the problems in the first place. In fact a result of carbon trading is that it has spawned yet another way for wheelers and dealers to make more money. Putting economists in charge of tackling climate change is tantamount to putting Dracula in charge of a blood-bank, or Tony Blair in charge of peace in the Middle East!

Even if the use of markets to tackle climate change did have an effect, it is important to note that the 'burden of adjustment', as economists refer to the pain that results from change, would fall almost exclusively on the shoulders of those least able to bear it. The poor would be effectively excluded from activities such as flying and driving, while the rich of course would continue merrily along their trajectory of conspicuous consumption.

The good news, such as it is, is that there are some economists, such as the New Economics Foundation in the UK, who seem to be aware of the issues that we have been discussing and are actively involved in seeking alternative solutions. These must inevitably involve a combination of:
- stricter and more effective regulation
- changes in the fundamental way in which we approach life and the planet, which brings us back to the basic issues we have been discussing: everything is connected
- international co-operation. The environment is the classic example of a global issue: it is pointless for countries to act in isolation over global warming. In this the developed countries find themselves in a moral dilemma: how can we ask China and India to approach growth in a different way, when it is us in the developed world that have caused the problem over the last century or two? Of course it is in the

interest of the developing world as well to tackle climate change, since they too will have their lives and their livelihoods disrupted (more so in the case of the poorest countries in Africa, which are likely to be worst affected). So the way forward is fraught with difficulty. An example of this is the shenanigans over Kyoto, which have clearly demonstrated the difficulties of acting in concert in this area.

Baruchello: Your worries, and your hopes, are mine. This is also why I have been attempting to spread further the knowledge of John McMurtry's work outside Canada. His philosophy is, straightforwardly, a philosophy of life, acknowledged and supported by the United Nations themselves. No other contemporary philosopher has spoken as adamantly as he has done against the ongoing ecological collapse and its direct relation to today's predominant economic-political ideology.

Through your work, perhaps, British and Continental economists may now start receiving some 'McMurtryan' input; whilst philosophers, through mine, especially but not exclusively in the Nordic and Mediterranean countries, might get a sense of what the New Economics Foundation stands for. More than anything else, I believe this chapter— our dialogue—should serve as a sign of how relevant insights on contemporary issues can transmigrate from a given area of inquiry to another, across disciplinary boundaries. As stated before, the mind is the place where we turn modifiable human arrangements into dogmas, cages and straightjackets; but it is also the place where we can be freed from them, disciplinary fields and fences included.

Bibliography

Charles Tandy (ed.), *Death and Anti-Death, Volume 5: Thirty years After Loren Eiseley (1907-1977)*, Palo Alto: Ria Press, 2007.

F. Brower, V. Lintner and N. Newman (eds.), *Democracy and the European Union*, Federal trust, 1994.

Endnotes

[1] Giorgio Baruchello, "Deadly Economics: Reflections on the Neoclassical Paradigm", *Death and Anti-Death, Volume 5: Thirty years After Loren Eiseley (1907-1977)*, edited by Charles Tandy, Palo Alto: Ria Press, 2007, pp. 65-132.

[2] See F. Brower, V. Lintner and N. Newman (eds.), *Democracy and the European Union*, Federal trust, 1994 -- in particular the chapter by V. Lintner.

[3] The quote at issue reads: 'Capitalism is the extraordinary belief that the nastiest of men for the nastiest of motives will somehow work for the benefit of all.'

[4] There is of course the added attraction of possible capital gain.

CHAPTER TWO

Charles Hartshorne

Dan Dombrowski*

Charles Hartshorne is considered by many philosophers
to be one of the most important philosophers of religion and
metaphysicians of the twentieth century. Although
Hartshorne often criticized the metaphysics of substance
found in medieval philosophy, he was very much like
medieval thinkers in developing a philosophy that was
theocentric. Throughout his career he defended the
rationality of theism and for several decades was almost
alone in doing so among English-language philosophers.
Hartshorne was also one of the thinkers responsible for the
rediscovery of St. Anselm's ontological argument. But his
most influential contribution to philosophical theism did not
concern arguments for the *existence* of God, but rather was
related to a theory of the *actuality* of God, i.e., *how* God
exists. In traditional or classical theism, God was seen as the
supreme, unchanging being, but in Hartshorne's process-
based or neoclassical conception, God is seen as supreme
becoming in which there is a factor of supreme being. That
is, we humans become for a while, whereas God *always*
becomes, Hartshorne maintains. The neoclassical view of
Hartshorne has influenced the way many philosophers
understand the concept of God. In fact, a small number of
scholars—some philosophers and some theologians—think
of him as the greatest metaphysician of the second half of the
twentieth century, yet, with a few exceptions to be treated
below, his work has not been very influential among analytic
philosophers who are theists.

1. Life

Charles Hartshorne was born in the nineteenth century and lived to philosophize in the twenty-first. He was born in Kittanning, Pennsylvania (U.S.A.) on June 5, 1897. He was, like Alfred North Whitehead, the son of an Anglican minister, although many of his ancestors were Quakers. After attending Haverford College he served in World War One in France as a medic, taking a box of philosophy books with him to the front. After the war Hartshorne received his doctorate in philosophy at Harvard, and there he met Whitehead. Most of the major elements of Hartshorne's philosophy were already apparent by the time he became familiar with Whitehead's thought, contrary to a popular misconception. From 1923-1925 a postdoctoral fellowship took him to Germany, where he had classes with both Husserl and Heidegger. But neither of these thinkers influenced his philosophy as much as C.S. Peirce, whose collected papers he edited with Paul Weiss. In addition to many visiting appointments, Hartshorne spent his teaching career at three institutions. From 1928-1955 he taught at the University of Chicago, where he was a dominant intellectual force in the School of Divinity, despite the fact that he was housed in the Philosophy Department, where he was not nearly as influential. He was at Emory University from 1955 until 1962, when he moved to the University of Texas at Austin. Hartshorne eventually became a long-term emeritus professor at Austin and lived there until his death on October 9, 2000. His wife, Dorothy, was as colorful as her husband and was mentioned often in his writings. Hartshorne never owned an automobile, nor did he smoke or drink alcohol or caffeine; he had a passion for birdsong and became an internationally known expert in the field.

2. Method

Three primary methodological devices or procedures are at work in Hartshorne's metaphysics. First, he very often uses a systematic exhaustion of theoretical options—or the development of position matrices, sometimes containing thirty-two alternatives (!)—in considering philosophical problems. This procedure is evident throughout his philosophy, but it is most apparent in his various treatments of the ontological argument. To take another example, he thinks it important to notice that regarding the mind-body problem there are three options available to us, not two, as is usually assumed: some form of dualism, some form of the materialist view that psyche is reducible to body, *and* some form of the panpsychist (or, as he terms it, psychicalist) view that body is in some way reducible to psyche if all concrete singulars (e.g., cells or electrons) in some way show signs of self-motion or activity. Thomas Nagel famously considers this third option, but Hartshorne actually defends it.

Second, Hartshorne frequently uses the history of philosophy to see which of the logically possible options made available by position matrices have been defended before so as to avail ourselves of the insights of others in the effort to examine in detail the consistency of these positions and to assess their consequences. Nonetheless, those logically possible options that have not historically found support should be analyzed both in terms of internal consistency and practical ramifications. It should be noted that Hartshorne's use of the history of philosophy often involves lesser known views of famous thinkers (like Plato's belief in God as the soul for the body of the whole natural world, or Leibniz's defense of panpsychism) as well as the thought of lesser known thinkers (such as Faustus Socinus, Nicholas Berdyaev, or Jules Lecquier).

Third, after a careful reading of the history of philosophy has facilitated the conceptual and pragmatic examination of all the available options made explicit by a position matrix, the (Greek) principle of moderation is used by Hartshorne as a guide to negotiate the way between extreme views on either side. For example, regarding the issue of personal identity, the view of Hume (and of Bertrand Russell at one stage in his career) is that, strictly speaking, there is no personal identity in that each event in "a person's life" is externally related to the others. This is just as disastrous, Hartshorne thinks, as Leibniz's view that all such events are internally related to the others, so that implicit in the fetus are all the experiences of the adult. (This Leibnizian view relies on the classical theistic, strong notion of omniscience, wherein God knows in minute detail and with absolute assurance what will happen in the future.) The Humean view fails to explain the continuity we experience in our lives and the Leibnizian view fails to explain the indeterminateness we experience when considering the future. The truth lies between these two extremes, Hartshorne thinks. His view of personal identity is based on a conception of time as asymmetrical in which later events in a person's life are internally related to former events, but they are externally related to those that follow, thus leading to a position that is at once partially deterministic and partially indeterministic. That is, the past supplies necessary but not sufficient conditions for human identity in the present, which always faces a partially indeterminate future.

Only the first of these methodological devices or procedures supports the widely held claim that Hartshorne is a rationalist. His overall method is a complex one that involves the other two methods or procedures, where he does borrow from the rationalists, but also from the pragmatists and the Greeks. It must be admitted, however, that Hartshorne was educated in a philosophic world still heavily influenced by late nineteenth and early twentieth century idealism.

3. The Existence and Actuality of God

Philosophers commonly use a metaphor that suggests that the chain of an argument, say for the existence of God, is only as strong as its weakest link. Hartshorne rejects this metaphor on Peircian grounds. He replaces it by suggesting that various arguments for the existence of God— ontological, cosmological, design, etc.—are like mutually reinforcing strands in a cable.

He argues that Hume's and Kant's criticisms of the ontological argument of St. Anselm are not directed at the strongest version of his argument found in *Proslogion*, chapter 3. Here, he thinks, there is a modal distinction implied between existing necessarily and existing contingently. Hartshorne's view is that existence alone might not be a real predicate, but existing necessarily certainly is. To say that something exists without the possibility of not existing is to say something significant about the being in question. That is, contra Kant and others, Hartshorne believes that there are necessary truths concerning existence. To say that absolute nonexistence in some fashion exists is to contradict oneself; hence he thinks that absolute nonexistence is unintelligible. It is necessarily the case that *something* exists, he thinks, and, relying on the ontological argument, he also thinks it necessarily true that God exists.

On Hartshorne's view, metaphysics does not deal with realities beyond the physical, but rather with those features of reality that are ubiquitous or that would exist in any possible world. And he does not think that it is possible to think of a preeminent being that only existed contingently since if it did exist contingently rather than necessarily, it would not be preeminent. That is, God's existence is either impossible (positivism) or possible, and, if possible, then necessary (theism). He is assuming here that there are three alternatives for us to consider: (1) God is impossible; (2) God is possible, but may or may not exist; (3) God exists necessarily. The ontological argument shows that the second alternative makes no sense. Hence, he thinks that the prime task for the philosophical theist is to show that God is not impossible.

In addition, Hartshorne's detailed treatment of the argument from design is connected to his view of biology. It is hard to reconcile an omnipotent, classical theistic God with all of the monstrosities and chance mutations produced in nature, but the general orderliness of the natural world is just as difficult to reconcile with there being no Orderer or Persuader at all. Belief in God as omnipotent, he thinks, has three problems: (1) it is at odds with the disorderliness in nature; (2) it yields the acutest form of the theodicy problem; and (3) it conflicts with the notion from Plato's *Sophist*, defended by Hartshorne, that being *is* dynamic power (*dynamis*). An *omni*potent being would ultimately have all power over others, who would ultimately be powerless. But any being-in-becoming, according to Hartshorne, has *some* power to affect, or to be affected by, others; this power, however slight, provides counterevidence to a belief in divine omnipotence. In contrast, God is ideally powerful, on the Hartshornian view. That is, God is as powerful as it is possible to be, given the partial freedom and power of creatures.

Hartshorne's dispute with traditional or classical philosophical theism concerns not so much the *existence* of God, but rather its assumption that the *actuality* of God (i.e., *how* God exists) could be described in the same terms as the existence of God. A God who exists necessarily is not necessary or unchanging in every other respect (e.g., in terms of divine responsiveness to creaturely changes), he thinks. Although Hartshorne believes that the medieval thinkers were correct in trying to think through the logic of perfection, he also thinks that this logic has traditionally been misapplied in the effort to articulate the attributes of a being called "God," roughly defined as the greatest conceivable being. The traditional or classical theistic logic of perfection sees God as monopolar in that regarding various contrasts (permanence-change, one-many, activity-passivity, etc.) the traditional or classical philosophical theist has chosen one element in each pair as a divine attribute (the former element of each pair) and denigrated the other.

By way of contrast, Hartshorne's logic of perfection is dipolar. Within each element of these pairs there are good features that should be attributed in the preeminent sense to God (e.g., excellent permanence in the sense of steadfastness, excellent change in the sense of preeminent ability to respond to the sufferings of creatures). In each element in these pairs there are also invidious features (e.g., pigheaded stubbornness, fickleness). The task for the philosophical theist, he thinks, is to attribute the excellences of both elements of these pairs to God and to eschew the invidious aspects of both elements. However, it should be noted that *some* contrasts are not fit for dipolar analysis (e.g., good-evil) in that "good good" is a redundancy and "evil good" is a contradiction. The greatest conceivable being, he thinks, cannot be evil in any sense whatsoever.

Hartshorne does not claim to believe in two gods, nor does he wish to defend a cosmological dualism. In fact, we can see that the opposite is the case when we consider that, in addition to calling his theism *dipolar*, he refers to it as a type of *panentheism*, which literally means that all is *in* the one God by means of omniscience (as Hartshorne defines the term) and omnibenevolence. All creaturely feelings, especially feelings of suffering, are included in the divine life. God is seen by Hartshorne as the mind or soul for the whole body of the natural world (see above regarding Plato's World Soul), although he thinks of God as distinguishable from the creatures. Another way to categorize Hartshorne's theism is to see it as *neoclassical* in the sense that he relies on the classical or traditional theistic proofs for the existence of God and on the classical theistic metaphysics of being as *first steps* in the effort to think through properly the logic of perfection. However, these efforts need to be supplemented, he thinks, by the efforts of those who see becoming as more inclusive than being. God is not outside of time, as in the Boethian view that is influential among traditional philosophical theists, but rather exists through all of time, on Hartshorne's view. On the neoclassical view, God's permanent "being" consists in steadfast benevolence exhibited through everlasting becoming.

God is omniscient, on Hartshorne's view, but "omniscience" here refers to the divine ability to know everything that is knowable: past actualities as already actualized; present realities to the extent that they are knowable according to the laws of physics (e.g., what is present epistemically may very well be the most recent past, given the speed of light); and future possibilities or probabilities *as possibilities or probabilities*. On the traditional or classical conception of omniscience, however,

God has knowledge of future possibilities or probabilities as already actualized. According to Hartshorne, this is not an example of supreme knowledge, but is rather an example of ignorance of the (at least partially) indeterminate character of the future.

The asymmetrical view of time, common to process thinkers in general (e.g., Bergson, Whitehead, Hartshorne), in which the relationship between the present and the past is radically different from the relationship between the present and the future, also has implications for Hartshorne's theodicy. A plurality of partially free agents, including nonhuman ones, facing a future that is neither completely determined nor foreknown in detail, makes it not only possible, but likely, that these agents will get in each other's way, clash, and cause each other to suffer. On this view, God is the fellow sufferer who understands.

4. Axiology

Hartshorne views the cosmos as a "metaphysical monarchy," with God as the presiding, but not omnipotent, head, and he sees human society as a "metaphysical democracy," with each member as an equal. This makes him a liberal in politics if "liberalism" refers to the egalitarian belief that none of us is God. Although Hartshorne and Whitehead are both political liberals, Hartshorne is, despite his view of panpsychist reality as thoroughly social, more of a libertarian liberal and Whitehead more of a redistributive liberal. In axiology as well as in metaphysics/theodicy, freedom is crucial, on Hartshorne's view.

Hartshorne's panpsychism (or psychicalism) entails the belief that each active singular in nature, even those like electrons and plant cells that do not exhibit mentality, is nonetheless a center of intrinsic, and not merely instrumental, value. As a result, Hartshorne's metaphysics is meant to provide a basis for both an aesthetic appreciation of the value in nature, as well as for an environmental ethics where intrinsic and instrumental values in nature are weighed.

As a published expert on bird song, Hartshorne is the first philosopher since Aristotle to be an expert in both metaphysics and ornithology. He writes specifically of the aesthetic categories required to explain why birds sing outside of mating season and when territory is not threatened—two occasions for bird song that are crucial to the behaviorists' account. Birds *like* to sing, he concludes.

Hartshorne's criticism of anthropocentrism is due not only to his concern for God, but also for beings-in-becoming who experience in a less sophisticated way than humans. To say that all active singulars feel—leaving out of the picture abstractions like "twoness" or insentient composites of active singulars that do not themselves feel as wholes—is not to say that they are self-conscious or that they think. As before, however, Hartshorne's axiology is ultimately theocentric in character.

5. Critical Evaluation

It seems fair to say that analytic philosophers, in general, even analytic philosophers who are theists, have largely ignored Hartshorne's philosophy. (The point is debateable. There has been a move among many analytic philosophers who are theists away from the eternal, Boethian God who is outside of time altogether. Might it be that Hartshorne's influence is greater than initially appears to be the case when the temporality, or the sempiternity, of the God of many analytic philosophers is concerned?) This is in contrast to his wide influence among theologians, which is odd when it is considered that he is not a theologian and does not rely on sacred scripture or religious authority for his insights. Another oddity is the fact that Hartshorne's influence among theologians is due to the defense he offers of the *rationality* of belief in a neoclassical God.

There is at least one important philosopher whose work indicates the sort of debate that has occurred between Hartshorne and analytic theists, who tend to rely on traditional or classical versions of the concept of God. That is William Alston. There are two reasons why an evaluation of Hartshorne's philosophy is facilitated by a consideration of Alston's critique. First, Alston is as important a theist as any among analytic philosophers and his criticisms of Hartshorne's thought are like those of other analytic philosophers like Thomas Morris, Richard Creel, Michael Durrant, Colin Gunton, and others. And second, Alston is a former student of Hartshorne's and is thoroughly familiar with Hartshorne's arguments. Alston is a philosopher who is not scandalized by Hartshorne's panentheism, nor by his neoclassical theism. But Alston thinks that the contrast that Hartshorne draws between his neoclassical theism and the classical theism of Thomas Aquinas is too sharp.

Alston thinks that Hartshorne presents neoclassical theism and classical theism as complete packages, whereas it would be better to be able to pick and choose among individual items within these packages. Alston seeks some sort of rapprochement between Thomism and neoclassical theism, a rapprochement that Hartshorne himself would like to bring about to the extent that he is a neo*classical* thinker, but that is difficult to accomplish to the extent that he is *neo*classical.

A consideration of ten contrasting attributes will best facilitate an initial understanding of Hartshorne's view of God. Consider [on the following page] the first group of attributes treated by Alston.

CLASSICAL ATTRIBUTES	NEOCLASSICAL ATTRIBUTES
1. absoluteness (absence of internal relatedness)	1. relativity (God is internally related to creatures by way of knowledge of them and actions toward them)
2. pure actuality (there is no potentiality in God)	2. potentiality (not everything is actualized that is possible for God)
3. total necessity (every truth about God is necessarily true)	3. necessity and contingency (God exists necessarily, but various things are true of God contingently, e.g., God's knowledge of what is contingent)
4. absolute simplicity	4. complexity

Alston distinguishes two lines of argument regarding absoluteness and relativity, which he sees as the key contrast. Alston thinks that only one of these is successful. As indicated in the diagram above, what Hartshorne means by absoluteness is absence of internal relatedness. A relation is internal to a term of a relation just in case that term would not be exactly as it is if it were not in that relationship. Hartshorne's view is that God has internal relations to creatures by way of knowing and acting towards them.

On Alston's interpretation, Hartshorne's first line of argument is to say that if the relation of the absolute to the world really fell outside the absolute, then this relation would necessarily fall within some further and genuinely single entity that embraced both the absolute and the world and the relations between them. Thus, we must hold, according to Hartshorne, that the God-creature relation is internal to God; otherwise we will have to admit that there is something greater or more inclusive than God. Alston does not find this argument convincing because it includes the claim that God "contains" the world due to the internal relations God has with the world. Alston's view is that the entity to which a relation is internal contains the terms only in the sense that those terms enter into a description of the entity, but it does not follow from this that those terms are contained in that entity as marbles are in a box.

Divine inclusiveness, for Hartshorne, is sometimes like the inclusion of thoughts in a mind, but usually it is described as like the inclusion of cells within a living body. It is never like the inclusion of marbles in a box. The inorganic and insentient character of a box is inadequate as a model for divinity, he thinks, and divine inclusiveness is never like the inclusion of theorems in a set of axioms, as it might be for certain idealists. Divine inclusiveness in Hartshorne is *organic* inclusiveness.

Hartshorne's second argument against absoluteness fares much better, according to Alston. He agrees with Hartshorne's stance regarding the cognitive relation God has with the world; in any case of knowledge, the knowledge relation is internal to the subject, external to the object. When a human being knows something, the fact that she knows it is part of what makes her the concrete being that she is. If she recognizes a certain tree she is different from the being she might have been if she had not recognized the tree. But the tree is unaffected by her recognition. Likewise, according to Alston, one cannot maintain that God has perfect knowledge of everything knowable and still hold that God is not qualified to any degree by relations with other beings.

One might respond to Alston and Hartshorne on this point by saying that since creatures depend for their existence on God, their relations to God affect *them*, but not God. Richard Creel seems to make this very point. But even if beings other than God depend for their existence on God, it still remains true that if God had created a different world from the one that exists at present, then God would be somewhat different from the way God is at present: God's knowledge would have been of *that* world and not this one, according to both Alston and Hartshorne.

Alston's concessions to Hartshorne's concept of God extend to contrasts 2-4. The above argument for the internal relatedness of God (as cognitive subject) to the world presupposes that there are alternative possibilities for God, and if there are alternative possibilities for divine knowledge then this implies that there are unrealized potentialities for God. *Pure* actuality and *total* necessity cannot be defended as divine attributes, according to Alston and Hartshorne. Alston's version of Hartshorne's argument goes as follows:

(1) (a) "God knows that W exists" entails (b) "W exists."

(2) If (a) were necessary, (b) would be necessary.

(3) But (b) is contingent.

(4) Hence (a) is contingent.

We can totally exclude contingency from God only by denying God any knowledge of anything contingent, a step that not even traditional or classical theists wish to take. Contrast 4 must also be treated in a dipolar way in that the main support for a doctrine of pure divine simplicity is the absence of any unrealized potentialities in God.

In sum, Alston and Hartshorne agree on contrasts 1-4, except for the fact that Hartshorne's concept of divine inclusiveness, in contrast to Alston's, is organic in character.

Regarding a second group of attributes, however, Alston and many other theists who are analytic philosophers diverge from Hartshorne rather significantly:

CLASSICAL ATTRIBUTES	NEOCLASSICAL ATTRIBUTES
5. creation *ex nihilo* by a free act of will; God could have refrained from creating anything	5. both God and the world of creatures exist necessarily, although the details are contingent
6. omnipotence (God has the power to do anything God wills to do that is logically consistent)	6. God has all the power one agent could have given metaphysical, in addition to logical, limitations
7. incorporeality	7. corporeality (the world is the body of God)
8. nontemporality (God does not live through a series of moments)	8. temporality (God lives through temporal succession, but everlastingly)
9. immutability (God cannot change because God is not temporally successive)	9. mutability (God is continually attaining richer syntheses of experience)
10. absolute perfection (God is eternally that than which no more perfect can be conceived)	10. relative perfection (at any moment God is more perfect than any other, but God is self-surpassing at a later stage of development)

Concerning contrast 5, Alston takes creation *ex nihilo* to be fundamental to theism because it has deep roots in religious experience. He thinks that to say that God has unrealized potentialities and contingent properties is not to say that God *must* be in relation with some world of entities other than God. Alston admits that Hartshorne legitimately points out some of the internal contradictions contained in the classical theistic version of creation *ex nihilo*, but he claims that there is no connection drawn by Hartshorne between divine creation and metaphysical principles regarding relativity, contingency, and potentiality. Alston's belief seems to be that those who accept creation *ex nihilo* are not saying that there is absolutely nothing at any stage: there is God. Rather, creation *ex nihilo* only means that there is nothing out of which God creates the universe. Here Alston seems to agree with Norman Kretzmann, Eleonore Stump, and most other theists who are analytic philosophers.

Alston's stance here is problematic for two reasons, from Hartshorne's point of view. First, although belief in *some* sort of divine creativity has deep roots in the history of religious experience, it is not clear that these roots have to tap into creation *ex nihilo*. For example, it is not clear that creation *ex nihilo* is the sort of creation described in Genesis, in that when the Bible starts with the statement that the spirit of God hovered above the waters, one gets the impression that both God and the aqueous muck had been around forever. If one believes in creation *ex nihilo*, however, as Alston does, one might nonetheless claim that creation *ex nihilo* does not necessarily mean a temporal beginning to the act of creation. But even on this hypothesis there are problems, and this would seem to be Hartshorne's second point. If Plato and Hartshorne are correct that being *is* dynamic power, then the sort of unlimited power implied by creation *ex nihilo* is impossible. Hartshorne would argue, contra Alston, that there is a connection between belief in creation *ex hyle* (as opposed to creation *ex nihilo*) and the metaphysical principle that being is dynamic power. Creation *ex nihilo*, Hartshorne thinks, is a convenient fiction invented in the first centuries B.C.E. and C.E. in order to exalt divine power, but it is not the only sort of creation that religious believers have defended, nor is it defensible if being is dynamic power.

Concerning contrast 6, Alston claims that belief in creation *ex nihilo* and belief in divine omnipotence are separate beliefs such that to argue against the former is not necessarily to argue against the latter. Hartshorne tries to do too much, he thinks, with the claim that being is power when he uses this claim to argue against divine omnipotence. According to Alston, God can have *unlimited* power, power to do anything that God wills to do, without having *all* power in that, if being is power, the creatures also have some power.

On Hartshorne's interpretation of Alston, however, God can have unlimited power, but not all power, because God delegates some power to others. Although God does not have all power, Hartshorne thinks that on Alston's view God *could* have all power. In effect, what Alston has done, according to Hartshorne, is reduce his stance regarding divine omnipotence to that regarding creation *ex nihilo* in that the claim that God could have all power is due to the prior belief that God brings everything into existence out of absolutely nothing, a belief that Alston thinks has to be the traditional one and in point of fact is intelligible. It is not quite clear to Hartshorne, however, that it is unquestionably the traditional one, nor is it clear to him that we can develop an intelligible concept of "absolutely nothing."

Hartshorne's Platonic or Bergsonian argument against creation *ex nihilo*, in simplified form, looks something like this: one can in fact imagine the nonexistence of this or that, or even of this or that class of things, a fact that gives some the confidence to (erroneously) think that this process can go on infinitely such that one could imagine a state in which there was "absolutely nothing." However, not every verbally possible statement is made conceptually cogent by even the most generous notion of "conceptual," according to

Hartshorne. At the specific, ordinary, empirical level negative instances are possible, but at the generic, metaphysical level only positive instances are possible, on this view. The sheer absence of reality cannot conceivably be experienced, he thinks, for if it were experienced an existing experiencer would be presupposed.

Contrast 7 deals with divine embodiment. Alston is willing to grant that God is embodied in two senses: (1) God is aware, with maximal immediacy, of what goes on in the world; and (2) God can directly affect what happens in the world. That is, Alston defends a limited version of divine embodiment, similar to that defended by Richard Swinburne. However, Alston is sceptical regarding a stronger version of divine embodiment wherein the world exists by metaphysical necessity such that God *must* animate it. Alston is willing to accept the idea that God has a body, but *only if* having such a body is on God's terms. It seems that this weaker version of divine embodiment defended by Alston, as opposed to Hartshorne's stronger version wherein there is essential corporeality in God, stands or falls with the defense of creation *ex nihilo*. In fact, despite Alston's desire to examine each contrast individually, as opposed to Hartshorne's stark contrast between classical theistic attributes (all ten of them) and neoclassical attributes (all ten of them), he ends up linking his criticisms of Hartshorne regarding contrasts 5-7, at the very least. All three of these classical theistic attributes stand together only with a defensible version of creation *ex nihilo*.

Contrasts 8-9, concerning nontemporality and immutability, are also linked by Alston. He concedes that if God is temporal, Hartshorne has offered us the best version to date of what divine temporality and divine mutability would be like. Alston dismisses as idle the view that God could remain completely unchanged through a succession of temporal moments, but this admission still leaves us, he thinks, with the following conditional statement: "God undergoes change *if* God is in time." Alston's critique of Hartshorne's view also consists in a refusal to grant that contingency and temporality are coextensive in the way mutability and temporality are. Alston believes, contra Hartshorne, that God can be in some way contingent (that any relation in which God stands to the world might have been otherwise) and still be nontemporal.

Alston knows that the notion of a nontemporal God who is qualified by relation to temporal beings will strike Hartshorne as unintelligible. Alston's attempt to make his position intelligible rests on his own Thomistic-Whiteheadian stance, or better, on his Thomistic or Boethian interpretation of Whitehead (strange as this seems). We should not think of God as involved in process or becoming of any sort. The best temporal analogy, he thinks, for this conception is an unextended instant or an "eternal now." For Alston this does not commit one, however, as Hartshorne would allege, to a static deity frozen in immobility. On the contrary, according to Alston, God is eternally active in ways that do not require temporal succession. God's acts can be complete in an instant. Alston includes God's acts of knowledge, a stance that at least seems to conflict with one of the concessions he made to Hartshorne regarding the first group of attributes.

The Boethian-Thomistic notion of the specious present for God, on the analogue of a human being's perceiving some temporally extended stretch of a process in one temporally indivisible act, is also defended by Alston. For example, one can perceive the flight of a bee "all at once" without first perceiving the first half of the stretch of flight and then perceiving the second. One's perception can be without temporal succession even if the object of one's perception is, in fact, temporally successive. All we have to do, on Alston's view, is expand our specious present to cover all of time and we have a model for God's awareness of the world. This is a much more difficult project for Hartshorne to imagine than it is for Alston. Apparently Alston thinks that it is easy to conceptualize God "seeing" Neanderthal man (or Adam), Moses, Jesus, and Dorothy Day all at once in their immediacy. Here Alston has a view similar to that of William Mann.

But even if it were possible to have nonsuccessive *awareness* of a vast succession, which Hartshorne would deny, it is even more implausible, from Hartshorne's point of view, to claim, as does Alston, that God could have nonsuccessive *responses* to stages of that succession. It might make more sense for Alston to say "indesponses" or "presponses" rather than "responses," as Creel would urge.

It is correct of Alston to notice that there is no loss in God, but this is not incompatible with God's temporality, according to Hartshorne. There can be succession in God without there being loss or perishing due to the fact that God's inheritance of what happens in the world and God's memory are ideal. Hartshorne thinks that the future is incomplete and indeterminate for God as well as from our limited perspective. Alston, by way of contrast, wants to defend a God who is not strictly necessary in actuality, but is contingent, *despite the fact that* God does not undergo temporal change, nor is God fluent. Hartshorne's defenders, by way of contrast, think that one of the greatest virtues of process thinking is its effort to eliminate what they see as such self-contradiction in philosophical theology.

Alston's treatment of contrast 10, concerning absolute versus relative perfection, follows from what he has said regarding contrasts 8-9. Relative perfection in God, as opposed to absolute perfection, has a point only for a temporal being; hence God is absolutely perfect, according to Alston. A being that does not successively assume different states could not possibly surpass itself. Here, once again, Alston engages in linkage, thereby, at the very least, confirming Hartshorne's belief that we need both to consider the divine attributes together and to determine whether the classical theist's linkage or the neoclassical theist's linkage is more defensible. For the most past, Alston opts for classical theism. Or more precisely, he thinks that the strongest concept of God is acquired when we take a modified version of the neoclassical attributes in contrasts 1-4 and combine them with the classical attributes in contrasts 5-10.

This rapproachement in Alston between classical theism and neoclassical theism is a step beyond James Ross's belief that these are two competing descriptions of God at an impasse. Hartshorne seems to agree with Ross. Neoclassical, dipolar theism *already* includes the best insights of classical theism, he thinks.

From Hartshorne's point of view the linkage of attributes *within* the first group and *within* the second group needs to be corrected by a greater concern for reticulating the attributes in these two groups. He thinks that an explanation is needed regarding how Alston can be committed to both monopolar and dipolar theism. For example, Alston ends up defending the view that God is changed by the objects God knows (pace the neoclassical, dipolar attributes), but these are not changes that occur in time (pace the classical, monopolar attributes). It is one thing, Hartshorne thinks, to say that God exists in a nontemporal specious present, and it is another to say that God is changed by temporal beings in a nontemporal specious present. The former view is at least problematic, he thinks, and the latter seems to be part of the traditional classical theistic view wherein, from a Hartshornian perspective, inconsistency goes in the guise of mystery.

Bibliography

Books by Hartshorne

- (1923) "An Outline and Defense of the Argument for the Unity of Being in the Absolute or Divine Good." Ph.D. Dissertation, Harvard University.
- (1934) *The Philosophy and Psychology of Sensation.* Chicago: University of Chicago Press.
- (1937) *Beyond Humanism.* Chicago: Willet, Clark, and Co.
- (1941) *Man's Vision of God.* N.Y.: Harper and Brothers.
- (1948) *The Divine Relativity.* New Haven: Yale University Press.
- (1953) *Reality as Social Process.* Boston: Beacon Press.
- (1953) *Philosophers Speak of God.* Chicago: University of Chicago Press.
- (1962) *The Logic of Perfection.* LaSalle, Il.: Open Court.
- (1967) *A Natural Theology for Our Time.* LaSalle, Il: Open Court.
- (1967) *Anselm's Discovery.* LaSalle, Il.: Open Court.
- 1970) *Creative Synthesis and Philosophic Method.* LaSalle, Il.: Open Court.
- (1972) *Whitehead's Philosophy.* Lincoln: University of Nebraska Press.
- (1973) *Born to Sing.* Bloomington: Indiana University Press.
- (1976) *Aquinas to Whitehead.* Milwaukee: Marquette University Press.
- (1983) *Insights and Oversights of Great Thinkers.* Albany: State University of New York Press.
- (1984) *Existence and Actuality: Conversations with Charles Hartshorne.* Chicago: University of Chicago Press.

- (1984) *Creativity in American Philosophy*. Albany: State University of New York Press.
- (1984) *Omnipotence and Other Theological Mistakes*. Albany: State University of New York Press.
- (1987) *Wisdom as Moderation*. Albany: State University of New York Press.
- (1990) *The Darkness and the Light*. Albany: State University of New York Press.
- (1991) *The Philosophy of Charles Hartshorne*. LaSalle, Il.: Open Court.
- (1997) *The Zero Fallacy and Other Essays in Neoclassical Metaphysics*. LaSalle, Il.: Open Court.

Secondary Sources

- Alston, William. (1964) "The Elucidation of Religious Statements." in *Process and Divinity: The Hartshorne Festschrift*. LaSalle, Il.: Open Court.
- ----- (1989) "Hartshorne and Aquinas: A Via Media." in *Divine Nature and Human Language*. Ithaca: Cornell University Press.
- Auxier, Randall, ed. (2001) *Hartshorne and Brightman on God, Process, and Persons*. Nashville: Vanderbilt University Press.
- Cobb, John and Franklin Gamwell, eds. (1984) *Existence and Actuality: Conversations with Charles Hartshorne*. Chicago: University of Chicago Press.
- Dombrowski, Daniel. (1988) *Hartshorne and the Metaphysics of Animal Rights*. Albany: State University of New York Press.
- ----- (1996) *Analytic Theism, Hartshorne, and the Concept of God*. Albany: State University of New York Press.
- ----- (2004) *Divine Beauty: The Aesthetics of Charles Hartshorne* . Nashville: Vanderbilt University Press.

- ----- (2006) *Rethinking the Ontological Argument: A Neoclassical Theistic Response.* New York: Cambridge University Press.
- Ford, Lewis, ed. (1973) *Two Process Philosophers.* Tallahassee, Fl.: American Academy of Religion.
- Gilroy, John. (1989) "Hartshorne and the Ultimate Issue in Metaphysics." *Process Studies* 18: 38-56.
- Goodwin, George. (1978) *The Ontological Argument of Charles Hartshorne.* Missoula: Scholars Press.
- Griffin, David Ray. (2001) *Reenchantment without Supernaturalism.* Ithaca: Cornell University Press.
- Gunton, Colin. (1978) *Becoming and Being: The Doctrine of God in Charles Hartshorne and Karl Barth.* Oxford: Oxford University Press.
- Hahn, Lewis, ed. (1991) *The Philosophy of Charles Hartshorne.* LaSalle, Il.: Open Court.
- Morris, Randall. (1991) *Process Philosophy and Political Ideology.* Albany: State University of New York Press.
- Peters, Eugene. (1970) *Hartshorne and Neoclassical Metaphysics.* Lincoln: University of Nebraska Press.
- Ross, James. (1977) "An Impasse on Competing Descriptions of God." *International Journal for Philosophy of Religion* 8: 233-249.
- Shields, George, ed. (2003) *Process and Analysis: Essays on Whitehead, Hartshorne, and the Analytic Tradition.* Albany: State University of New York Press.
- Sia, Santiago, ed. (1990) *Charles Hartshorne's Concept of God.* Boston: Kluwer.
- Towne, Edgar. (1997) *Two Types of New Theism: Knowledge of God in the Thought of Paul Tillich and Charles Hartshorne.* New York: Peter Lang.
- Tracy, David. (1985) "Analogy, Metaphor, and God-Language: Charles Hartshorne." *Modern Schoolman* 62: 249-265.

- Viney, Don. (1985) *Charles Hartshorne and the Existence of God.* Albany: State University of New York Press.
- ----- (1998) "The Varieties of Theism and the Openness of God: Charles Hartshorne and Free-Will Theism." *Personalist Forum* 14: 196-234.
- Vitali, Theodore. (1977) "The Peircian Influence on Hartshorne's Subjectivism." *Process Studies* 7: 238-249.
- Whitney, Barry. (1985) *Evil and the Process God.* Toronto: Edwin Mellon Press.

Other Resources

- The Center for Process Studies, at the Claremont School of Theology

---

* Reprinted by permission from:

Dombrowski, Dan, "Charles Hartshorne", *The Stanford Encyclopedia of Philosophy (Winter 2008 Edition),* Edward N. Zalta (ed.). URL = <http://plato.stanford.edu/archives/win2008/entries/hartshorne/>.

AFTERWORD

There is an important addendum that should be mentioned. It concerns Hartshorne's view of death, which mediates between two extremes. One extreme is the nontheistic view that death ends all. The other is the classical theistic view that human beings survive death in terms of the personal immortality of the soul. As a theist, Hartshorne rejects the former. But as a neoclassical theist he also rejects the latter in that he thinks both that the dualism of mind and matter on which personal immortality is based is problematic and that belief in personal immortality is an indication of a certain hubris on the part of human beings.

Whereas Kurt Gödel famously considered, and perhaps flirted with, the ontological argument for the existence of God, Hartshorne explicitly defends a modal version of the argument. One implication of this defense is that whereas God exists necessarily, we exist only contingently. As an ornithologist Hartshorne was greatly influenced by the fact that human beings exhibit the limitations of all biological organisms, including death. To claim necessary or immortal existence for human beings is to suggest that they are, or deserve to be, divine, which Hartshorne sees as idolatrous.

His own view is called "contributionism." We are vicariously immortal in the sense that our experiences and accomplishments in life are contributions to an omniscient and omnibenevolent God who will everlastingly remember and cherish them. From Hartshorne's point of view, what it means to be religious is to receive solace from the realization that such contributions find an everlasting home. Death ends personal existence, but it does not end all. We die, but God lives on everlastingly.

CHAPTER THREE

Choosing Death In Cases Of Anorexia Nervosa – Should We Ever Let People Die From Anorexia?
PART II

Simona Giordana

1. INTRODUCTION [1]

This paper is a continuation of the arguments proposed in my 'Choosing Death In Cases Of Anorexia Nervosa – Should We Ever Let People Die From Anorexia?' published in Volume 5 of *Death and Anti-Death*. As discussed in that chapter, a large part of the debate around the right to refuse life-prolonging treatment for anorexia nervosa centers on competence. The general assumption, accepted in medical ethics and protected by law, is that competent refusals of medical treatment must be respected. As we have seen in Volume 5, anorexics could be competent to refuse life-saving treatment. However, I argued that the anorexic's refusal of life-saving treatment should not be respected purely because it is a competent decision. Anorexia has two characteristics that weaken the normative strength of the principle of respect for competence. Firstly, anorexia is not a terminal illness; secondly, it involves the family in a profound way. Whereas in Volume 5 I explained how the most often cited 'Incompetence Arguments' are mistaken, in this chapter I discuss how reversibility of anorexia and family participation might weaken the normative strength of competence, and the implications of this.

2. COMPETENCE

Competence is regarded as central to clinical decisions both in medical ethics and in medical law. Competence is the standard by which many clinical decisions are made: if a patient is competently refusing medical treatment, her decision ought to be respected. If not respected, the treating doctor would violate bodily integrity, thus exposing her/himself to charges for assault or battery (under English law). When the decision is not competently expressed, it can be overridden in the patient's best interests.

Competence, thus, has crucial normative strength: it dictates what courses of action ought or can be taken. One exception to this rule, in the UK, is the treatment of people sectioned under the Mental Health Act 1983. Section 63 of the Act establishes that consent to treatment for the illness for which the patient is hospitalized shall not be requested. But, as I have argued elsewhere [2] , this exception is not ethically sound, as people who are compulsorily hospitalized because of a mental illness can be competent to make decisions relating to their illness. It is not clear why those who are compulsorily hospitalized by reason of their mental illness should not be treated with the same respect as those who are not hospitalized compulsorily.

Anorexia typically affects intelligent, young and well-educated people. Many of those with eating disorders are talented and highly functional. They are very different from the stereotype of the 'insane', detached from reality and incapable of appreciation of the meaning of their choices. Think, for example, of Princess Diana, Whitney Huston, Alanis Morrisette, Anna Freud, the model Beverly Johnson, the actor Billy Bob Thorton, the singer Karen Carpenter, the gymnast Christy Hienrich [3] . They have all been affected by eating disorders, but could hardly be described as insane or incompetent at directing their life. Any medical intervention against the wishes of someone who is intelligent, lucid and capable of clear understanding of the consequences of their decisions can be much harder to justify than interventions against the stated wishes of patients who are detached from reality or who have severe intellectual impairments (unless one wanted to argue that coercion is justified when people have a mental illness, an argument that makes no sense, as we have seen in Volume 5).

Whereas competence adds moral strength to a refusal of treatment, competence should not be the sole test, in decisions relating to suspension or omission of life-prolonging treatment for anorexia. Refusal of treatment could be ethically overridden (temporarily), even if it is a competent choice, if the anorexic stands a good chance of recovery. Such refusal could also be ethically overridden (temporarily), to allow the meaningful others to come to terms with the patient's death. I am not proposing that competence should be abridged entirely, but that it should be tempered. Let us now see how reversibility in anorexia is normatively significant. Later, we shall see why the family can legitimately participate in the therapeutic decisions.

3. CONTROVERSIES AROUND DEATH: A CASE HISTORY

The bottles of water portrayed on the following page are the symbol of a protest that has split public opinion in Italy for the best part of a decade. Hundreds of bottles have been layed in Milan, under the Dome, and at the front door of the clinic that hosts Eluana Englaro. Eluana is a PVS patient since 1992.

Her father, who is her legal guardian, requested suspension of therapy when it was clear that she was not going to recover. This wish was partly based on the irreversibility of her condition. Partly, it was based on Eluana's known wishes. Before she fell ill, she said she would not wish to be kept alive in a vegetative state. The father's request has for long been rejected: under Italian legislation refusing nutrition is a *diritto personalissimo*, a 'very personal right', which cannot be exercised on behalf of somebody else. In October 2007, the *Corte di Cassazione* began to ventilate the possibility of a resolution to this case. Since then, the Roman Catholic Church began a strong opposition to the Italian magistrates. The Church requested that Eluana be handed over to them. They would nourish the vulnerable patient, sentenced by the Italian State to a horrible death by starvation and dehydration through no fault of her own. Public opinion was split, with large numbers walking to

Milan, in pilgrimages of salvation, and laying symbolic bottles of water at the door of the clinic where Eluana is still being kept alive against her previously stated wishes [4].

Other cases (Diane Pretty [5], Piergiorgio Welby [6] just to mention some) have given rise to controversies both in ethics and law and in public opinion. All these cases illustrate that, whereas in principle the right to refuse medical treatment is uncontested, in practice decisions to exercise that right can generate acrimonious controversies, according to the modalities in which death has to be brought about and to the circumstances in which patients opt for death. There is nothing approaching a peaceful consensus concerning decisions to dispose of our life.

One thing that emerges from these cases is that irreversibility and intractability of suffering have moral significance. People who stood up for Eluana's father, and, previously, for Piergiorgio Welby, did so not only because they had competently stated the wish to be allowed to die, but also because their condition was irreversible, and there was no hope to alleviate their suffering.

In anorexia, the whole secondary symptomatology is reversible, and the vast majority of people who develop eating disorders recover. Some might resolve entirely the conflicts and anxieties relating to the body. Others, instead, might accept a healthy weight, but continue to experience insecurity over the body, or to negotiate with themselves over what is acceptable to eat. Yet, in spite of the different meanings of 'recovery' for different individuals, the vast majority of anorexia sufferers neither die by starvation, nor commit suicide [7]. A person with anorexia can be helped to develop a good relationship with self and to get over the anxieties and fears that generate anorexia.

4. LEGAL AND MORAL RELEVANCE OF REVERSIBILITY

The proposal of temporarily overriding a competent refusal of treatment, if the anorexic is treatable, is apparently a departure from accepted ethico-legal principles. A competent refusal of treatment, *prima facie*, ought to be respected even if the treatment promises a full cure of the disease. However, this proposal ('of temporarily overriding...if...treatable') is consistent with legislation and seems to incorporate the moral feelings of the majorities.

For example, in Oregon, one of the first places to legalize euthanasia, the law allows euthanasia *only for terminally ill patients*. In The Netherlands, the terminality of the illness is not a legal requirement. But in order to obtain euthanasia, the suffering has to be unbearable and the treatment hopeless. Suffering can be either mental or physical (House of Lord 2005 a, S. 171 and S. 210, The Netherlands).

In the UK, the Assisted Dying Bill, proposed in 2005 reads:

> *Qualifying conditions [for physician assisted suicide]*
> *(2) (b) the patient has a terminal illness; and (d) the patient is suffering unbearably as a result of that terminal illness* (House of Lords 2005 b).

In 2007, The World Federation of Right to Die Societies Newsletter, discussing the proposal to legalize euthanasia in South Australia, laid before the House of Assembly in May, stated:

> *The essential elements for legislation [on euthanasia] are that the condition is irremediable by medical treatment and the suffering is intolerable to the patient* [8]

Invariably, one crucial justification for shortening someone's life is the intractability of the condition and the suffering of the patient. Several surveys show that public opinion also regards euthanasia acceptable in cases of untreatable illnesses or of patients in intractable pain [9].

One could object that these further qualifying conditions about irreversibility and intractability should only stand for euthanasia and assisted suicide, which require an active intervention of a third party to shorten someone's life. Instead, the right to refuse medical treatment should be granted only on the basis of competence.

However, refusal of treatment and euthanasia are, in an important way, morally equivalent. What matters morally in both cases is whether a person should be allowed to die prematurely, when her or his life could be prolonged. Issues of whether others have a moral obligation to do something to help us to hasten our death, or issues about the ways in which it could be ethical to bring about death, are additional moral questions. The central moral question is whether someone whose life could be saved (in anorexia, saved easily) should be allowed to die [10].

Competence should not cloud all other considerations. Whether or not a person is suffering unbearably, whether or not that suffering can be alleviated, are morally relevant facts. A request for suspension of treatment, albeit competent, is morally more difficult to sustain, when the illness is entirely reversible and the patient, albeit in despair at a certain point in time, has significant chances to recover and might probably be thankful for being rescued; it is reported that many anorexics who have been rescued against their wishes will thank you for that [11] .

5. ONE OBJECTION

One might object that patients have the right to refuse medical treatment even if the illness that affects them is treatable. Should, for example, Jehovah's Witnesses be denied the right to refuse whole blood products? Why should the anorexic be forced to receive life saving fat and the Jehovah's Witness not forced to receive life saving blood?

There are morally relevant differences between anorexia and other treatable illnesses, which justify the difference in treatment that I am proposing.

Preventing death in anorexia might give the patient the chance to survive and, some time down the line, to recover. A modification of beliefs and priorities is inherent to recovery from anorexia. The fears and anxieties that lead the person to starvation and other abuses of the body can be modified and abandoned.

Curing the illness of the Jehovah's Witness will not modify his beliefs and priorities, because his illness and his religious beliefs operate on two different psychological plans. Curing the illness has nothing to do with changing the beliefs that support refusal of medical treatment. Treating anorexia, instead, aims precisely at changing the psychological dynamics that lead to refusal of medical treatment.

One could object here that I am effectively suggesting that there are defects in competence in anorexia [12] . If a condition is only treatable provided that the person's beliefs and priorities change, this seems to imply that the person has defects in competence. The Jehovah's Witness in my example does not show such defects, and this explains why he and the anorexic should be treated differently. However, the preferences of the anorexic might be lethal without necessarily resulting from defects in competence. And yet, in order to get better, the person has to change those preferences. Of course one can always raise doubts about people's genuine autonomy – but these can be extended to the Jehovah's Witness, who effectively is willing to die because he believes that, by taking foreign blood, he will go to hell. The soundness of that belief can also be questioned. Yet, both the anorexic and the Jehovah's Witness can have the relevant competence to refuse medical treatment. People can have harmful preferences that are not necessarily an indication of defects in competence.

I am not suggesting that we should force anorexics to eat, but that refusal of treatment in anorexia is not as straightforward as refusal of life saving treatments for other conditions, and that a difference in treatment can be ethically justified in light of the difference in the condition.

6. THE RIGHT TO BE WRONG

At the 9th World Congress of Bioethics, organized by the International Association of Bioethics in Croatia, in September 2008, after a presentation of this argument, it was objected to me that that I am proposing to give up the *right to be wrong*, one of the most precious liberties we have. Doctors should not be entitled to force us to be 'right': indeed at common law the patient has the right to be wrong as long as she has the required understanding [13].

I however wonder what we should do with the right to be wrong, when the 'mistake' makes us dead. Mistakes that make us dead are not like all other mistakes, as we cannot regret them. The fact that there is no remedy to death seems a good enough reason to try to prevent people from making 'mistakes' that will lead them to death, when it is not clear that they want to die, or when they might be helped to flourish and enjoy life again. We can have a second chance to respect people's wish to die, whereas they will not have a second chance to live, once they 'got it wrong', and died as a result.

We now see in what ways and why the family could also advance ethically legitimate claims to direct the therapeutic decisions relating to the anorexic, and what is the extent of their entitlement.

7. THE MORAL PLACE OF THE FAMILY

One implication of the principle of respect for competence is that the therapeutic process is set between two poles: doctor or medical team and patient. This relationship is protected by privacy and confidentiality.

The family (more properly, the significant others) acquires a moral or legal entitlement to participate in the therapeutic process either if the patient is incompetent (in this case, under English law, they can give information relating to what the patient would have wanted, if competent, whereas in other jurisdictions their role can be more marginal – see the case of Eluana, mentioned at the beginning of this paper), or when the patient involves one of more family members in the therapeutic relationship. With these exceptions, normally family members cannot consent to treatment, or refuse treatment, or request treatment, on behalf of a competent patient.

In many cases of anorexia, the family should instead *prima facie* be involved in decisions relating to therapy, especially in final decisions on whether a patient should be allowed to die. There are at least **three** reasons why the family should be involved in end of life decisions relating to the anorexic:

First, anorexia is not just an intra-psychic condition. According to some experts, anorexia is a 'systemic condition' [14]. Turner, for example, talks about 'the anorexic family', rather than about the 'anorexic person' [15]. The condition typically appears when the patient is a minor, and it is often the family who refers the sufferer to medical services. Family dynamics change as a result of the condition. Families often devote an enormous amount of resources, including financial resources, in the attempt to rescue the sufferer from the tragic grip of eating disorders [16]. When a patient gets to the stage of receiving life-prolonging therapy, this is in many cases after years of fights that have involved the whole family. In virtue of their engagement, and of the emotional, physical and financial resources devoted to the anorexic, the family acquires a moral entitlement to have their sentiments considered [17]. This entitlement is of course not absolute. When the family's wishes cause significant disadvantage or even harm to the patient, those wishes should not be respected. If the family is never going to accept the patient's choices, this will of course not be a reason to keep the patient alive indefinitely against her wishes. However, where this is not the case, the family who has participated directly in the therapeutic process should not be excluded at the end of it, in the name of 'respect for competence'.

Secondly, involvement of the family can enhance the patient's competence. Understanding the impact of refusal of treatment on the close relatives can increase the patient's capacity to make an informed judgment over life-prolonging therapy. Consider the following case history. Mrs. Black (not her real name) was a "44-year-old married female with two children, a son aged 20 and a daughter aged 17. Mrs. Black had a 25-year history of anorexia nervosa. At the time of the evaluation she suffered from anorexia nervosa and related medical complications. She was living in a hospice connected to the hospital at which she had received much of her treatment over the years, admitted by her physician who felt she might have less than 6 months to live [...] The referral for an evaluation was generated when Mrs. Black requested the option to refuse life supports the next time her medical condition deteriorated. In 1997, she had previously been twice on life support systems. Following these episodes she decided that she would refuse such measures in the future. Her husband, at the time her legal guardian, was 'in agreement with her intention and requests an evaluation to be certain that [his wife] is competent to make this decision'." [18]

In assessing Mrs. Black's competence, the ethics committee posed, among others, the following questions:

> Could she understand the imminence of her death?

> To what extent was she aware of the effects of her anorexia on herself and her family?

The assessment of Mrs. Black's competence revealed that she had not fully understood the impact of her choices on her children. "She spoke most clearly about its effect on her mother; she was least clear about its effect on her daughter. She described her mother as being 'devastated...she'll be lonely and guilty.' [...] Her daughter, she thought, would be 'sad,' but Mrs. Black could not elaborate. She gave a more nuanced account of its effect on her son and was tearful discussing her husband, saying '[It's been] 21 years and I still don't know [how he feels].' [...] The conflict with her husband overshadowed her ability to think about saying good-bye to him" [19].

In this case, the medical team and the ethics committee organized meetings with the whole family, including the patient, and discussed with all of them the decision to omit further life-prolonging treatment. This enabled the sufferer to make a more informed choice about dying. Competence is understood as the capacity to appreciate the consequences of one's choices. These consequences should include the impact on family members in cases in which the family has participated actively in the treatment of the anorexic. In this way involvement of the family might increase the patient's capacity to refuse life-prolonging treatment.

Finally, involvement of the family can promote the patient's welfare. In the case of Mrs. Black, for example, assessment also revealed that conflicts with her husband were preventing her from parting from all of them on peaceful terms. Her inability to say goodbye to the children was also due to the fact that the conflicts with her husband were overshadowing everything else. The husband also recognized that he wished he could say goodbye without tensions. Counseling allowed them to elaborate these conflicts, and Mrs. Black to depart without unresolved acrimonies. This process not only benefited the patient, who died without the anguish of having unresolved frictions with any of her closest relatives, but also the survivors, who had the possibility to come to terms with the decision in a more serene way. A process that can balance the interests of all parties, and protects the welfare of all those involved, is surely to be preferred to one that does not.

The extent of the family's entitlements has to be balanced against the interests of the patient. Medicine should not of course serve the interests of the family to the detriment of the interests of the patient. Where the family's wishes could only be respected at the cost of causing significant harm or disadvantage to the patient, the best interests of the patient should take precedence. The family should not in fact be allowed to *override* the patient's competent refusal of treatment. It should simply be allowed the opportunity to elaborate regarding the patient's death, and this should be granted, in principle, even if this means that the patient's requests are not honored immediately. In reality, this is the way in which most of these decisions are made. In a number of reported cases of anorexics who have been let die, families and ethics committees have deliberated together on whether the patient's requests should be satisfied [20]. The principle of respect for competence does not explain why this is an ethical way of handling the situation. A mitigated principle of respect for competence leaves space to the sentiments and needs of the caring others.

8. NON-VOLUNTARY EUTHANASIA FOR ANOREXICS

When a patient with anorexia requests suspension of therapy, or omission of life-prolonging therapies, how the medical team should respond to these requests should not only be based on competence, but also on the appreciation of the futility of treatment and the moral place that significant others might have. If the patient's request is beyond any reasonable doubt competent, this adds further, and important, moral strength to her/his request. However, even if some doubts might remain on the patient's competence (for example, because of prolonged starvation), this is not in itself a reason to exclude suspension of life-prolonging therapy. If medical treatment is futile, and the patient suffers intractable pain, request for suspension or omission of life-prolonging therapy and administration of palliation should be considered. Medical treatment might be futile both for the competent and incompetent.

In addition to this, as it is pointlessly cruel to let people suffer more when they could suffer less, if the patient whose refusal of treatment was respected further requested help with hastening her/his death, there would be no ethical reason to deny it. Indeed, in The Netherlands in 1994, an anorexic patient was given a lethal drug. [21]

One could object that if the incompetent should be allowed to die on request, this could justify non-voluntary euthanasia. Indeed, this is a proper inference. By 'nonvoluntary euthanasia' it is generally meant the administration of a lethal drug ('active' euthanasia), or the omission/suspension of life-saving or life-prolonging treatment ('passive' euthanasia), to a patient who is not competent to make or express a wish on the matter. Indeed, allowing an anorexic who does not appear beyond any reasonable doubt competent to refuse life-prolonging treatment or to receive euthanasia might be properly called 'nonvoluntary euthanasia'. But nonvoluntary euthanasia is not necessarily unethical or repugnant. First, nonvoluntary euthanasia is not murder: it is not the killing of someone who wants to live. Here we have patients who state their wishes clearly, and state them repeatedly over a period of time. Secondly, nonvoluntary euthanasia is standard medical practice. People who are fully and permanently incompetent are routinely allowed to die, if available treatment is futile, in the sense of being unable to redeliver them back to a dignifying life. Suspension or omission of life-prolonging treatment is permitted in the patient's best interests: for example, in cases of irreversibly comatose patients [22], or patients with severe brain damage. This indicates that withdrawing or withholding treatment of those who are not competent is not necessarily repugnant. If it is ethically acceptable to suspend life-prolonging treatment in cases in which patients can in no way express their views, it must surely be acceptable to do the same when conscious, lucid, intelligent people in touch with reality reiterate their request to have that treatment suspended, even if not all doubts about their competence can be cast aside.

Finally, it is clear that someone who is at 'death's door' might well have diminished capacities. But if we took this as a justification for coercive treatment, this could potentially disintegrate people's entitlement to refuse medical treatment. The macabre paradox would be the following: the anorexic would be allowed to competently refuse food, with a full entitlement to non-interference based on competence. But when starvation is taking the toll on her, she would be ethically force-fed, up to the point of regaining minimal competence to enable her to competently decide again to engage in extreme dieting. When, once more, at death's door, she could be ethically force-fed again, being discharged once she had regained sufficient capacity to get back to starving.

The argument that the dying patient should be forcibly treated because s/he is not competent to die, thus, is surely nonsense and has undesirable consequences. If we want to force an anorexic to live, it is much better to do so before she is about to die.

9. CONCLUSIONS

Whether or not an anorexic should be allowed to refuse life-saving treatment does not depend solely or primarily on competence. It also depends on whether suffering can be alleviated, on whether the condition is treatable and on the type of involvement of the family. I have argued that time and care should be devoted to those who have been involved in the fight against anorexia. If the patient stands a good chance to recover, a competent request for omission or suspension of medical treatment could be ethically, albeit temporarily, overridden. If treatment is futile, and if the patient's suffering is intractable, her request should be considered, even if not all doubts on her competence can be eliminated.

Whereas in principle it is always possible to recover from anorexia, a minority of patients will not recover. In Volume 5 I have given indication of when treatment can be regarded as futile. When treatment is futile, patients should be allowed to go, and when this decision is made, they should be allowed to die in the most peaceful way, not purely because (or if) they are competent, but because medicine should not force any human being into pointless agony. Medicine should not condemn people into unnecessary suffering, not even in the name of some allegedly ethical principles, such as the 'Value of Life', or of similar others, which often fill some principled moral circles with reverberant echoes, when the living person can no longer assign any value to her/his own life, and only wishes to have her torments suspended.

BIBLIOGRAPHY

Airedale NHS Trust v Bland [1993] 1 All ER 281.

Amarasiekara Kumar and Bagaric Mirko. 2004. Moving from Voluntary Euthanasia to Non-Voluntary Euthanasia: Equality and Compassion. *Ratio Juris* 17, no. 3:398-423.

De Haan J. 2006. The ethics of Euthanasia: Advocates' Perspectives. *Bioethics* 16, no. 2:154-72.

Donnison D. 1996. In *British Social Attitudes. The 13th Report,* ed. R. Jowell Dartmouth. http://www.euthanasia.cc/97-1dvd.html.

Gans M. and Gunn W. B. 2003. End stage anorexia: criteria for competence to refuse treatment. *International Journal of Law and Psychiatry* 26, no. 6:677-95.

Gargett E. 2001. Changing the Law in South Australia. *World Right-to-Die Newsletter*: 3 (The World Right-to-Die Newsletter is a publication of the World Federation of Right to Die Societies).

Giordano Simona 2003. Anorexia Nervosa And Refusal Of Naso-Gastric Treatment. A response to Heather Draper. *Bioethics*, 17(3):261-78.

Giordano S. 2005. *Understanding Eating Disorders*. Oxford: Oxford University Press.

Giordano S. 2005. Is the body a republic? Ethics of organ and tissue post- mortem retention and use. *Journal of Medical Ethics* 31:470-475.

Griffiths R. and Russell J. 1998. Compulsory Treatment of Anorexia Nervosa Patients. In *Treating Eating Disorders: Ethical, Legal and Personal Issues,* ed. W. Vandereycken and P. J. V. Beumont, 127. New York: New York University Press.

House of Lords 2005 a. Selected Committee, Assisted Dying for the Terminally Ill Bill, First Report, 3 March 2005, S.171 and 210. Available online at http://www.publications.parliament.uk/pa/ld200405/ldselect/l dasdy/86/8602.htm.

House of Lords. 2005 b. *Assisted Dying for the Terminally Ill Bill* [HL]© Parliamentary copyright House of Lords PUBLISHED BY AUTHORITY OF THE HOUSE OF LORDS. LONDON — THE STATIONERY OFFICE LIMITED, November 2005, The Stationery Office Limited.

Marino I. 2007. (President of the Health Commission of the Italian Senate), for example on http://www.radioradicale.it/il-dibattito-pubblico-sul-caso-welby.

Re J (A mior) (Wardship: Medical Treatment) [1990] 3 All ER 930 per Lord Donaldson MR at 938 and Bland per Lord Goff at 868. Quoted in Grubb A. 2001. Euthanasia in England – a Law Lacking Compassion? *European Journal of Health Law* 8:89-93, 90.

Rodriguez v A-G British Columbia (1994) 85 CCC (3 d) 15, Cory J. Quoted in Amarasiekara Kumar and Bagaric Mirko. 2004. Moving from Voluntary Euthanasia to Non-Voluntary Euthanasia: Equality and Compassion. *Ratio Juris* 17, no. 3:398-423.

Selvini Palazzoli M. Cirillo S. Selvini M. Sorrentino A. M. 1998. *Ragazze anoressiche e bulimiche: La terapia familiare.* Milano: Cortina.

Shapiro J. P. 2001 Euthanasia's Home, *US News*, 30 March. Available at www.globalaging.org/elderrights/us/euthanasia.htm.

South Australian Voluntary Euthanasia Society at http://www.saves.asn.au/resources/facts/fs02.php .

Treasure J. L. 1998. Anorexia and Bulimia Nervosa. In *Seminars in General Adult Psychiatry,* ed. G. Stein and G. Wilkinson, 858–902. London: Royal College of Psychiatrists.

Turner B. S. 1984. *The body and society: explorations in social theory*. Oxford: Blackwell.

Vandereycken W. 1999. Viewpoint: Whose competence Should We Question?. *European Eating Disorders Review* 6:1–3.

Watson T. L., Bowers W. A. Bowers and Andersen A. E. 2000. Involuntary Treatment of Eating Disorders. *American Journal of Psychiatry* 157:1806–1810.

ENDNOTES

1. An earlier version of this paper has been accepted for publication. Giordano Simona 2008. Anorexia and refusal of life saving treatment. The moral place of competence, suffering and the family. *Philosophy, Psychology and Psychiatry*, in press.

2. Giordano Simona 2005. *Understanding Eating Disorders*. Oxford: Oxford University Press, Oxford, 2005: Chapter 11.

3. Giordano Simona, *Moral Choices, Eating Disorders and Fitness, Ethical and Legal Aspects*, Routledge, London, forthcoming.

4. Giordano Simona 2008. Anorexia and refusal of life saving treatment. The moral place of competence, suffering and the family. *Philosophy, Psychology and Psychiatry*, in press.

5. BBC 12 May 2002, at http://news.bbc.co.uk/1/hi/health/1983457.stm

6. Newspaper *Il Tempo*, 22-12-2006. available at http://www.iltempo.it/politica/2006/12/22/222667-utanasia_attiva_contro_legge.shtml.

7. Giordano Simona 2003. Anorexia Nervosa And Refusal Of Naso-Gastric Treatment. A response to Heather Draper. *Bioethics*, 17(3):261-78.

8. Gargett E. 2001. Changing the Law in South Australia. *World Right-to-Die Newsletter*: 3 (The World Right-to-Die Newsletter is a publication of the World Federation of Right to Die Societies).

9. Donnison D., *British Social Attitudes. The 13th Report,* ed. R. Jowell Dartmouth, 1996, at http://www.euthanasia.cc/97-1dvd.html.

10. Though they are not treated as equivalent by UK law, because UK law makes a distinction between acts and omissions.

11. Watson T. L., Bowers W. A. Bowers and Andersen A. E. 2000. Involuntary Treatment of Eating Disorders. *American Journal of Psychiatry* 157:1806–1810.

12. I owe this observation to John Harris.

13. *Hopp v Lepp* [1979] 98 DLR (3d) 464 at 470 per J. Prowse.

14. Selvini Palazzoli M. Cirillo S. Selvini M. Sorrentino A. M. 1998. *Ragazze anoressiche e bulimiche: La terapia familiare.* Milano: Cortina: 22-23,122.

15. Turner B. S. 1984. *The body and society: explorations in social theory.* Oxford: Blackwell:192.

16. Selvini Palazzoli M. Cirillo S. Selvini M. Sorrentino A. M. 1998. *Ragazze anoressiche e bulimiche: La terapia familiare.* Milano: Cortina: 115-18.

17. Giordano S. 2005. Is the body a republic? Ethics of organ and tissue post- mortem retention and use. *Journal of Medical Ethics* 31:470-475.

18. Gans M. and Gunn W. B. 2003. End stage anorexia: criteria for competence to refuse treatment. *International Journal of Law and Psychiatry* 26, no. 6:677-95.

19. Gans M. and Gunn W. B. 2003. End stage anorexia: criteria for competence to refuse treatment. *International Journal of Law and Psychiatry* 26, no. 6:677-95.

20. Giordano Simona 2005. *Understanding Eating Disorders*. Oxford: Oxford University Press, Oxford, 2005: Chapter 11 and 13.

21. Shapiro J. P. 2001 Euthanasia's Home, *US News*, 30 March. Available at www. globalaging.org/elderrights/us/eut hanasia.htm.

22. Airedale NHS Trust v Bland.

CHAPTER FOUR

The Ethics Of Enhancement

Bill Grote and William Grey

As well as seeking *longer* lives we also seek *better* lives. Indeed living longer in the absence of good life quality would be a largely pointless pursuit—a point developed powerfully by Aldous Huxley in his dystopian allegory *After Many a Summer*. [1] As well as exploring ethical issues in *life extension* it is therefore of interest to explore ethical issues in *life enhancement*.

Recent advances in enhancement therapies based on psychoactive drugs, genetic engineering and brain prostheses have stimulated much debate amongst bioethicists. [2] This debate is complicated by ambiguities in the term "enhancement" and an associated vagueness in the therapy-enhancement distinction. The emerging ethical questions include concerns about fairness, equity, the realization of excellence and the importance of the means of its achievement. There is also disquiet about the possibility of an attenuation of the concept of the self, and a degradation of human dignity and respect. In addition, there are worries about the use of mechanistic cures or palliatives for complex problems of mind and brain, and the dangers posed by Promethean aspirations. There is also a more general concern that the use of neuro-enhancing drugs will be a forerunner of brain enhancement based on genetic engineering, and a danger that this may soften us up for a regimented "post-human" future, reminiscent of another of Huxley's powerful dystopian visions, set out in *Brave New World*. [3] Debate on these issues concerns not only present but also the potential use of therapeutic and brain enhancing agents.

In this paper we address the nature of enhancement and the problem of the therapy-enhancement distinction, and then critically review some of the arguments that deal with moral concerns arising from an application of these mind affecting interventions. We also explore whether there are separate ethical consideration concerning enhancement with psychoactive drugs and enhancement using genetic engineering. We consider also the possibility that the use of pharmacological enhancements may camouflage the risks posed by genetic change.

From psychoactive drugs to genetic engineering: a slippery slope?

Some bioethicists suggest that the use of psychoactive drugs is a step towards the general acceptance of neuro-genetic engineering, and that these procedures, by eroding human dignity and degrading human nature, could pave the way towards a disturbing "post-human" future. [4] Acceptance of a seemingly benign mind-affecting technology, it is suggested, might soften us up for more dangerous enhancement techniques which might be introduced later.

These concerns may however be alarmist. There are a number of reasons, in particular in connection with continuing advances in drug development, which will probably prevent such a situation from developing. For example, in the future psychoactive drugs may be tailored in their pharmacological effect and their dose may be adjusted to an individual's genetically determined response profile, thus having greater efficacy with fewer side effects. [5] Better-targeted psychoactive drugs administered under medical supervision may provide many of the benefits promised by genetic interventions, and it may be that these benefits can be provided sooner, more flexibly, more cheaply, more rapidly and reversibly, thereby obviating the need for genetic modification.

Brain enhancing drugs (to be considered below) generally operate by transiently increasing neurotransmitter levels in nerve synapses (that is, in the gaps separating neurons), thereby modulating or increasing neural activity and the formation of new synaptic connections. These changes may produce calming, mood brightening, wakefulness and learning and memory effects. By contrast, genetic engineering techniques—if applied to somatic and, more controversially, to germ cells—pose substantial ethical challenges that go beyond the use of psychoactive drugs.

Genetic changes may irreversibly program brain cells to produce altered or novel molecules or structures, or create switches that amplify or modulate behavioural dispositions. Moreover if germ cells are modified, there is the possibility that these changes may be passed on to offspring who had no part in making decisions that directly affect them. Altered gene sequences in reproductive cells may also give rise to unintended, and possibly long delayed, consequences to individuals resulting from the reshuffling of genes which occurs during reproduction.

For most people the concern about the transition from drugs to genetically engineered brain enhancement has little relevance. It is unlikely that many individuals will be able to obtain these psychoactive agents, let alone the foreshadowed genetic treatments. Most people are protected by their poverty from the harms—as well from the possible benefits—that these technologies present. [6]

We suggest that discussion of the ethics of the two types of intervention should be carefully distinguished, because although they share the common feature of brain enhancement, there are significant differences between the methods. Consequently, even in countries afflicted with the ills of affluence the argument about a "softening up" process, or an inexorable progression from one sort of enhancement to the next, is far from obvious. These technologies are, to a large degree, independent and the different methods of enhancement they embody need to be considered on their merits.

Enhancement, enhancers and moral considerations

A broad definition of cognitive enhancement refers to any chemical, electronic or procedural interventions that may increase or improve cognitive performance beyond the normal or average range. The definition of "normal" however is problematic and must take into account cultural factors, such as age, sex and ethnicity.

Some discussions of this topic employ a very broad definition of enhancement, including, for example, the use of prostheses, IVF treatment, insulin, glasses, artificial hips, telephones, private schools and coaching colleges. [7] Such a broad definition generates confusion by grouping together "enhancements" which are altogether different in function, location and effect. A hip replacement has no direct effect on brain function—apart from pain reduction and the benefit of increased exercise on brain circulation—and schools and coaching colleges enhance the brain through learning and memory formation that utilise time-honoured and uncontroversial methods. However even if discussion is limited to direct brain enhancements, there is still plenty of room for confusion because of the many unrelated ways the brain and mind may be enhanced, and the different effects produced by these interventions.

There is a perfectly natural type of brain enhancement which occurs as the brain grows in complexity in response to a complex and rich environment. This process evolved over eons through "niche construction" through which humanity has influenced its own natural selection and thereby the evolution of succeeding generations. In this way, a dynamic, self-reinforcing process of enhancement has generated ever greater individual and social complexity, through multi-layered processes involving genes, the brain, environment and culture. [8]

These dynamic developments are part of a continuing process of gradual and incremental (if unintended) neuro-enhancement. In addition, with increasing knowledge and technological advances, humanity has discovered (or invented) methods such as meditation, and deployed a number of psychoactive substances that range from the innocuous to the dangerous and addictive—legal and illegal. [9] These agents modulate and affect cognitive performance and thereby increase relaxation, tranquillity, wakefulness, euphoria, stamina, pleasure, feelings of well-being, happiness or physical performance, as well as reducing pain, fear and inhibitions. Such psychoactive agents were (and are still) also used in rituals which purport to communicate with spirits, gods, or ancestors. [10]

Most cultures make use of at least some of the mind-affecting substances listed above in patterns of usage which are often related to social advantage. This raises ethical questions however which are not much considered in the literature dealing with the "new" psychoactive drugs. Instead the authors typically focus on gene changing recombinant DNA technology, and newly developed psychoactive drugs. [11] Any concerns about threats to human dignity or human nature are projected into the future, thereby suggesting that our present situation is unproblematic.

It is not clear why the—socially and individually—damaging and costly neuro-enhancing agents like alcohol, nicotine, or heroin should be excluded from the discussion. These agents may have a greater potential for addiction and for more serious physical and psychological damage than, for example, selective serotonin reuptake inhibitors (SSRIs) such as Prozac or stimulants such as Ritalin, which certainly affect mood, character and consciousness. If rationality and autonomy contribute to our sense of self and self-worth, thereby contributing to our worthiness of respect, then insofar as new psychoactive drugs represent a threat to these constituents of our humanity, they thereby constitute a threat to human well-being. In order to understand the nature of the threat posed by these new drugs, it is important to understand how and where they act. [12]

Many of the psychoactive drugs now in use became available during the last four to five decades. Earlier versions of these antidepressant and antipsychotic drugs—such as Lithium, the Tricyclics or Monoamide Oxidase Inhibitors (MAOIs)—had too many side effects, or few enhancing effects over and above therapeutic effects, for them to have been used regularly for the enhancement of "normal" individuals.

The new psychoactive drugs were serendipitously discovered to have a specific therapeutic effect, usually that of increasing the levels of neurotransmitters in the synapses and synapse formation. [13] Elevated levels of specific neurotransmitters can have mood brightening, relaxing and calming effects, as well as affecting the formation, modification, retention and retrieval of memory. [14]

Depression and anxiety seem to be related to synaptic levels of serotonin, dopamine, acetylcholine or norepinphrine in circuits used for learning and memory. [15] Throughout life traits and memory can be added to, modified or lost through a continuing process of learning, memory formation and forgetting.

Psychoactive drugs such as Prozac or Ritalin increase and rebalance neurotransmitter levels and thereby increase nerve activity and synaptic formation. These drugs modulate neuron circuits that store memory, make us feel depressed, happy, anxious or relaxed, and thus modify factors which are products of our genetic traits as well as of our learning and experience.

Therapy and enhancement users

There are three overlapping groups in society seeking therapy or enhancement with psychoactive drugs. First there are the seriously impaired who are clearly in need of therapy, and whose lives may be a misery perhaps to the point of being suicidal. Such individuals are unable to function normally or productively in society or in school and college. There is no question that this group may be helped greatly by psychoactive drugs and that any costs to society or to themselves posed by possible side effects are outweighed by the gains of the treatment. This group clearly stands to gain from existing and yet to be developed psychoactive drugs.

There is a second group of mildly afflicted individuals who may suffer from mild depression, melancholy, stress, fear, anxiety, sadness, underperformance, disappointment or unhappiness. Also included in this group are children and adolescents whose behaviour may be boisterous or unfocussed. Some may believe that they have personality traits, perhaps a result of early psychological trauma, which they are not comfortable with, despite the fact that they may seem to function normally in society, and who consequently believe they might benefit from drugs like Prozac [16] or Ritalin [17] . There is often disagreement about whether these individuals deserve pharmacological treatment or whether they are in a predicament from which they should be able to extricate themselves, perhaps with the help of counselling and life style changes.

The third group consists of more or less normal individuals—though there is no sharp boundary which separates members of this group from the second—who seek a quick fix to their problems (and who may not want to exert themselves), as well as those with an eye to any easy advantage and who seek to learn faster, remember more, and stay calm and alert in exams, or in public performances. These individuals want to perform better, have more intense experience and perhaps smooth out life's cycles of unhappiness and happiness. The use of psychoactive drugs by this group purely for enhancement, or for the treatment of mild conditions, may be a cause for concern quite apart from medical side effects and possible psychological harm. This category of users raises questions about distributive justice, equity and fairness, as well as more profound ethical questions about our shared humanity. Is the non-therapeutic use of psychoactive drugs for enhancement justified in these cases? There are important differences between the use of psychoactive drugs for enhancement—to change mental capacity, personality, consciousness, or memory—and therapeutic uses of these agents.

Therapy versus enhancement

Therapeutic use includes treatments that remedy illness, deficiencies and disorders or restore function to a normal range. This use falls within the traditional domain of medicine. It includes treatment of conditions that are typically covered through national health schemes or private health insurance. Scarce health dollars tend to be restricted to treatments characterised as therapy, normally corrective procedures based on medical need, and often excludes various elective procedures.

An *enhancement* is a treatment that aims to extend function or performance beyond the natural, average or normal range. Its aim is to improve physical or psychological—and in some cases social—abilities, behaviours, functions or capacities. The "snorting" of crushed Ritalin—which at higher concentrations has similar effects to methamphetamine—is an example of psychoactive drug use for the purpose of psychological enhancement. A student suffering mild depression who takes SSRIs, and as a result becomes more focussed and relaxed, perhaps performing better during examinations, would be harder to categorise and perhaps may be considered to have received both therapy and enhancement.

The therapy-enhancement distinction fails to provide a clear boundary between acceptable and non-acceptable uses of psychoactive agents, and indeed other biotechnologies that are currently developing. In many situations "normality", or what has been described as species-normal function, may not exist, or be subject to change in a way that makes it unhelpful or meaningless as a goal of treatment. Therapy and enhancement are not mutually exclusive and the therapy-enhancement distinction is therefore unhelpful for differentiating between acceptable and non-acceptable uses of psychoactive agents.

The therapy-enhancement distinction is further complicated also because of the plasticity and responsiveness of the brain and the often elastic criteria used in diagnosis. Research indicates that various types of brain/mind disorders can be accompanied by a different shape, volume or level of activity of certain brain regions compared to the normal, or by differences in observed brain metabolism. [18] Once these disorders are better understood, we may be better able to make this distinction.

Elastic definitions of health, for example that of the United States National Institute of Health: "if a condition causes unhappiness, psychological pain and social disadvantage then it represents a disease and interventions to remedy it should be considered cures", [19] or that of the World Health Organisation: "health is a state of complete physical, mental and social wellbeing", [20] provide generous interpretations that will not be welcomed by those responsible for health budgets, and would include most of the individuals classified above in group two as eligible for therapy.

In practice, the vagueness and elasticity of the therapy-enhancement distinction means that it often comes down to what a medical practitioner says it is. The relationship then depends on the medical practitioner's responsiveness to "life or wellness" problems, and their perceptions of hypochondria or deception. Are they willing to consider examination nervousness, restlessness, mild depression, anxiety, fear, mild memory loss, lack of assertiveness, shyness, or work stress a problem worth treating? And are they willing—or do they have time?—to listen to and offer help to the distraught parent with the hyperactive child?

Ethical concerns

Ethical concerns about the use of psychoactive agents focus on questions of fairness and equity, the importance of means (natural and authentic versus artificial and furtive) in attaining a goal, and whether acceptance of such means might diminish standards and degrade social values. However the social consensus against the use of enhancing agents in competitive sport has not (yet) developed in the domain of cognitive endeavour. There are also concerns about the legitimacy of offering palliative cures for complex problems involving the mind, the self and free will, and worries about the erosion of human dignity and respect—vital human values. In addition there are less sharply focused concerns about the hubris of Promethean aspirations associated with more ambitious life-enhancing proposals—including the transhumanist proposal for radical longevity enhancement. An ethical evaluation of these issues might reasonably begin by considering the general (and familiar) bioethical principles of beneficence, nonmaleficence, respect for persons, and justice. [21]

The use of psychoactive drugs raises questions of fairness and equity that are poignant and subtle. A wealthy family providing their children with access to current (or future) enhancers will be adding a further dimension to existing social advantage—which may include already a privileged, stable and rich environment, attendance at well-resourced schools and colleges, and the opportunity to forge the right social connections for a successful future professional or business career. Subsequently, social and peer pressure might pressure more and more people to use enhancers, and the increase in the use of enhancers will inevitably diminish their advantage to users, while increasing the disadvantage for less privileged non-users while shifting the 'normal range' of the bell curve in the direction of improved performance. This will create a new 'normal' or 'average' for that endeavour, thereby redefining standards of normality.

There is a deeply ingrained expectation in society—aligned perhaps with the protestant work ethic—that people should work hard for what they obtain and that valuable lessons of life are learned through perseverance, discipline, hardship failure, and hard-won success. Although taking an occasional psychoactive drug to overcome shyness, or to stay calm in an important examination may seem acceptable, using drugs to win a game of bridge or for better sports performance seems questionable.

Parens talks of the importance of *means* in achieving an outcome. [22] He contrasts a situation where a child is given Ritalin to be quiet and pay attention, with a child that is taught to sit quietly and to follow a lesson. In the latter case the child has worked, made an effort and might be proud of its achievement. For the child on Ritalin taking the drug has diminished the experience; it has learned that one strategy for meeting the challenges of life is popping pills. The child might apply this approach to other problems on to adult life. Similarly the use of enhancers threatens our conception of excellence, when through hard work together with giftedness or creativity excellent works are produced. The use of enhancers diminishes the merit of the achievement of the outstanding craftsman, sports person, student, artist or professional. [23]

Root Wolpe points out that the argument that if normal attention or cognition is good then increased attention or cognition is better, has surprising consequences. [24] The brain processes information, provides emotional content and filters information inputs and memory outputs. A memory enhancer would need to be discerning. For example, one would not normally wish to remember the details of staring for one hour at the advertising in a bus shelter or be too burdened by memories of horrifying accidents or violent films. In performing this filtering and winnowing the brain ensures that learning and accessing memory becomes manageable and that emotional highs and lows do not become disabling. [25]

Carl Elliott asks rhetorically: "spiritual emptiness the search for self, alienation in the midst of abundance: are there traits any more American than these?" [26] In considering the negative aspects of one very popular treatment for this existential angst he writes:

Prozac treats the self rather than the disease ... it alters personality and feeds dangerously into the American obsession with competition and worldly success...and so ... it offers a mechanistic cure for spiritual problems... and ...for all the good they [these drugs] do, the ills that they treat are part and parcel of the lonely, forgetful and unbearably sad place where we live.

Elliott is disturbed by the "medicalization of unhappiness" and warns of the "tyranny of happiness". He recommends non-pharmacological solutions to existential problems and suggests that we need alternative approaches which focus on changing customs, vigilance and a tightening of diagnostic procedures. However this is problematic. While some suffering may be beneficial, drawing a line between productive and unproductive suffering is problematic. At what point should people be left to struggle with, and at what point should they be extricated from, their predicaments? Some individuals who are otherwise performing well may need a psychoactive agent to bring them into a normal range in stressful situations—such as sitting examinations or performing in public. There is no simple answer as to whether drugs should be prohibited, restricted or promoted and subsidised. There is a need, rather, to consider costs and benefits on a case by case basis while exercising a precautionary approach. Caution is clearly prudent in the light of the fact that many drugs now restricted or banned were once promoted as unproblematic panaceas, and whose addictive or destructive character only became apparent gradually.

Should those who are mildly affected by psychological afflictions—such as unhappiness, shyness, hypochondria, melancholia, or mild depression—but who are still functioning, if sub-optimally, be denied psychotropic drug treatment because of concerns about the effects of enhancement? Elliott, Fukuyama, Kass and other ethicists would recommend limiting psychotropic drug use and seeking alternative ways of addressing a wide range of social problems, some of which may be generated by the empty spaces left by our usurped gods.

There is little hope that the trends of individualism, hedonism, competitive aggression, and alienation will correct themselves spontaneously. That is, if our growth-addicted, market-driven consumer society is ultimately responsible for the angst-ridden "low serotonin society", then it may be that—in the absence of an improbably social and economic *volte-face*—the only remedy for many may be psychoactive drug palliation.

Human brains have been shaped by millions of years of evolution which have fine-tuned them as instruments for survival and reproduction. Interfering with these complex systems with psychoactive drugs should not be undertaken lightly. Attempting to reshape our natural dispositions and responses for whatever purposes represents, according to Kass *et al*, a "failure to respect the giftedness of the natural world." Such hubristic use of psychoactive drugs may reduce our ability "to act freely, for ourselves, by our own efforts and to consider ourselves responsible, worthy of praise or blame..." and lead us to "... a time of wilfulness over giftedness, of dominion over reverence." [27]

If this precautionary approach is justified with respect to the issue of cognitive enhancement—because we risk disturbing a hard-won balance to which we are adapted—it will apply even more to the issue of longevity enhancement through the use of drugs or by genetic engineering.

The respect which we owe to nature however is clearly not unconditional. Nature, after all, delivers not just exquisitely adapted biological systems, but pestilence such as smallpox, malaria, cancer, Alzheimers and AIDS. Devastating catastrophes are perfectly *natural,* and the use of diagnostic scans, antibiotics, vaccines, contraceptive pills and embryo screening are the product of human contrivance rather than nature. The sublime and mysterious power of nature—as well as its pitiless indifference—has slowly yielded to human inquiry and to our constantly expanding knowledge. Nature should be treated with respect—biological processes have evolved over billion-year geological time scales and have thereby proved their robustness—but not always with reverence and humility. While it is important to be mindful of the danger of Promethean hubris generating mishap or disaster—which may generate natural and catastrophic self correction—as Kass *et al* warn us, humility has its own dangers and *failing to act* may also have catastrophic consequences.

We have noted already that there is a wide and disparate variety of psychotropic substances and cognitive enhancement procedures in use—including meditation, alcohol, cannabis, coaching colleges, and caffeine. All of these may be used legitimately, at least in moderation, without apparent ill-effects. Why then should we not use the expanded psychopharmacopia even more widely, to redress existing social inequalities, and further enrich the human condition? [28]

A major concern about actual and potential use of psychoactive drugs—and genetic engineering—is that we may be introducing changes which reshape natural systems at a rate and on a scale which are quite different from anything that has been experienced in the course of human evolution. It may be that these changes are too rapid, too discontinuous and too chaotic for the correction of problems that will almost certainly arise and which we cannot yet see.

There are, then, clear dangers in making a wide range of psychoactive drugs freely available, even (and especially) if costs were low and side effects minimal. A society whose members are insulated from emotional upheaval and the pain and struggle of the human condition would probably be far from healthy and perhaps be at risk of fragmentation and social collapse. [29]

We suggest that the ubiquitous use of psychoactive drugs is neither desirable nor inevitable. Humanity's long and chequered experience with psychoactive drugs provides grounds for caution. The efficacy, benefits, side effects, and costs of the new enhancers need to be considered, and compared carefully with a range of benign alternatives. However the opportunity for (legal) multi-billion dollar profits, the possibility of selling enhancements together with soft addiction, and the widespread desire to find a panacea for life's troubling discomforts will almost certainly guarantee that the development, refinement, production and use of these drugs in affluent western countries will continue apace. [30]

The use of psychoactive drugs for non-therapeutic (or marginally therapeutic) purposes in the second and third groups identified above warrants continuing critical attention. We have noted a strong body of bioethical opinion counselling caution about rapidly expanding the use of these agents beyond clearly therapeutic cases. The debate about the wider use of psychoactive drugs is lopsided, with cautious counselling mixed with advice from unbridled enthusiasts, such as the so-called "paradise engineers". The debate is complex and polarised and, because psychoactive agents have the potential to radically reshape our cognitive and affective structures, urgent. We believe that it is important to avoid the extreme positions of the alarmists and the "paradise engineers", and recommend continuing with vigorous and searching examination of not just pharmacology but also of the salient ethical, cultural and socio-economic factors. We believe that these considerations apply both to the issue of cognitive enhancement and to the issue of longevity enhancement.

Conclusion

In many applications it seems that psychoactive drugs are used therapeutically and virtually certain that this use will increase further. Discussion of psychoactive drug use should be separated from genetic enhancement technologies, in particular those aimed at germ line changes. [31]

On balance and at our present level of biotechnology, it seems unlikely that human nature and consciousness are under threat by psychoactive drugs used for either therapeutic or enhancement purposes. However we need to remain vigilant over whatever blandishments or "devil's bargain" propositions that smart marketers might conjure up. [32]

It is important therefore to maintain a cautious, critical and comprehensive watch on developing enhancement technologies and to foster a continuing social debate involving bioethicists about their application and utilisation. Based on the concerns discussed in this paper we believe that the development and application of enhancement technologies—for cognitive enhancement, as well as for longevity—is problematic, especially with respect to applications involving germ line genetic engineering, and that their development should be carefully guided by a precautionary approach. [33]

Bibliography

Begley, D.J. *et al.* (eds). 2000. *The Blood Brain Barrier and Drug Delivery to the CNS*. New York: Dekker

Breggin, P.R. 2000. *Reclaiming Our Children: A Healing Solution for a Nation in Crisis*. Cambridge MA: Perseus

Breggin, P.R. 2004. Suicidality, violence and mania caused by selective serotonin reuptake inhibitors (SSRIs): A review and analysis, *International Journal of Risk and Safety in Medicine* 16: 31-49

Caldicott, F. 1998. Mental Disorders and Genetics: The Ethical Context, *Nuffield Council on Bioethics* report. URL: http://www.nuffieldbioethics.org/fileLibrary/pdf/mentaldisor ders2001.pdf

Caplan, A. 2003. 'Is Better Best?' *Scientific American*, Sep: 84-85

Caplan, A. 2003a. 'No-Brainer: Can We Cope with the Ethical Ramifications of New Knowledge of the Human Brain?' in Markus (2003): 95-106

Dawkins, R. 1999. *The Extended Phenotype*. Oxford: Oxford University Press.

Dennett, D. C. 1983. 'Information, Technology and the Virtues of Ignorance', *Daedalus* 112: 135-153.

Diller, L. 2002. 'Prescription Stimulant Use in American Children: Ethical Issues.' Presentation at the President's Council on Bioethics, Washington, D.C., 2 December. URL: http://www.bioethics.gov/transcripts/dec02/session3.html

Elliott, C. 1998. 'The Tyranny of Happiness: Ethics and Cosmetic Psychopharmacology', in Parens (1998): 177-188

Elliott, C. 2000. 'Pursued by Happiness and Beaten Senseless: Prozac and the American Dream', *Hastings Center Report* 30, No 2: 7-12

Fukuyama, F. 2002. *Our Posthuman Future: Consequences of the Biotechnology Revolution*. London: Profile Books

Grey, W. 1996. 'The Ethics of Human Genetic Engineering', *Australian Biologist*, 9: 50-56

Grey, W. 2005. 'Design Constraints for the Post-human Future', *Monash Bioethics Review* 24: 10-19

Huxley, A. 1939. *After Many a Summer*. London: Chatto & Windus

Huxley, A. 1932. *Brave New World*. London: Chatto & Windus

James, O. 1998. *Britain on the Couch: Treating a Low Serotonin Society*. London: Century.

Kass, L.R. 2003. *Beyond Therapy: Biotechnology and the Pursuit of Happiness*. New York: Regan Books.

Kass, L.R. 2003a. *Life, Liberty and the Defense of Dignity*. San Francisco: Encounter Books

Kramer, P.D. 1993. *Listening to Prozac*, New York: Penguin

Kramer, P.D. 2002. 'Happiness and Sadness: Depression and the Pharmacological Elevation of Mood.' Presentation at the President's Council on Bioethics, Washington, D.C., 12 September. URL:
http://www.bioethics.gov/transcripts/sep02/session3.html

Laland, K.N. *et al.* 1999. 'Niche Construction, Biological Evolution and Cultural Change'. *Behavioral and Brain Sciences* 23 (2001) URL:
http://www.bbsonline.org/documents/a/00/00/05/28/

LeDoux, J. 2003. 'The Self: Clues from the Brain' In LeDoux, J. *et al.* (eds) *The Self: From Soul To Brain.* Annals of the New York Academy of Sciences, Vol 1001.

Markus, S.J. (ed). 2002. *Neuroethics: Mapping The Field.* New York: Dana Press

NYAS 2003. 'Meeting Neuroethical Challenges in Cognitive Enhancement'. *Meeting of the New York Academy of Sciences*, 16-17 June. URL: http://www.nyas.org./ebriefreps/splash.asp?intebriefID=214

Parens, E. 1998. *Enhancing Human Traits: Ethical and Social Implications.* Washington, D.C., Georgetown University Press

Parens, E. 1998a. 'Is Better Always Good? The Enhancement Project'. In Parens (1998): 1-28

Parens, E. 2002. 'How Far Will the Treatment/Enhancement Distinction Get Us as We Grapple with New Ways to Shape Ourselves?' in Markus (2002): 152-158

Pinker, S. 2003. 'The Designer Baby Myth', *Guardian*, Thursday June 5

Restak, R. 2000. *Mysteries of the Mind.* Washington DC: National Geographic Society

Root Wolpe. P. 2002. 'Treatment, Enhancement, and the Ethics of Neurotherapeutics, *Brain and Cognition*, 50: 387-395

Root Wolpe, P. 2002a. 'Neurotechnology, Cyborgs, and the Sense of Self', in Markus (2002): 159-191

Root Wolpe, P. 2003. Social and Ethical Barriers to Enhancement. NYAC (2003)

Rothman, S., and Rothman D. 2003. *The Pursuit of Perfection: The Promise and Perils of Medical Enhancement*. New York: Pantheon

Salleh, A. 2003. 'Brain shrinkage: early sign of schizophrenia?' *ABC News in Science*, Aug 18. URL http://www.abc.net.au/science/articles/2003/08/18/925547.ht m

Scientific American 2003. 'The Brain Issue', September.

Singer, E. 2004. 'The Master Switch', *New Scientist*, Mar 6: 35-37

Sowell, E.R. *et al.* 2003. 'Cortical abnormalities in children and adolescents with attention deficit hyperactivity disorder'. *Lancet* 362 (9397): 1699-707

Studwell, J. 2004. Oh, behave, *Financial Times*, Jan 23. URL: www.FT.com

Endnotes

1. Huxley (1939).

2. See Kass (2003); Markus (2002); NYAS (2003); *Scientific American* (2003); Studwell (2004).

3. Huxley (1932).

4. Fukuyama (2002).

5. Two major advances which could transform treatments are changes affecting the master neurotransmitter glutamate (Singer 2004), and the development of methods to overcome the blood-brain barrier, which limits psychoactive drugs to lipophilic small molecules (Begley 2000). Overcoming the blood-brain barrier opens the possibility of creating a suite of new and better-targeted drugs.

6. Members of the third world are not in general afflicted by the alienation, fear, uncertainty, stress, loneliness, competition (as well as by the associated mindless profit- and consumption-driven behaviour) that generates the anxiety, melancholy and depression so widespread in western countries. This is not to deny that the less affluent majority have their own problems, as well as their own low-cost enhancements and remedies. The conditions of anxiety, fear, shyness, stress, aggression, restlessness and depression in the west have been described as a type of "synaptic sickness" (LeDoux 2003) whose sufferers constitute what has been called the "low serotonin society" (James 1998). This condition (if real) seems to be peculiar to wealthy western countries peopled by the sedentary, hedonistic and narcissistic, "me, I, myself" generation, with its culture of aggressive and competitive individuality—frequently accompanied by family breakdown, and elastic moral values.

7. Caplan (2003; 2003a).

8. Dawkins (1999); Laland (1999).

9. Mind affecting substances used—and abused—include: caffeine, alcohol (ethanol), nicotine, hashish, opium, kava, kat, coca leaf, betel nut, St Johns wort, ginko, brami, various inhalants, cactus or fungal extracts (e.g. LSD), cocaine, heroin and amphetamine derivatives.

10. There are also substances such as petrol, aerosol propellants or paint thinners often used by the young and the disadvantaged, that have destructive and debilitating effects.

11. Like the selective serotonin reuptake inhibitors (SSRIs) such as Prozac (fluoxitine-hydrochloride), or stimulants such as Ritalin (methylphenidate).

12. Restak (2000).

13. Synaptic connections constitute the complex network of approximately 100 trillion nerve synapses which form the systems that encode memory, the self, personality and character.

14. Ritalin and Adderall (amphetamine) are chemically related groups of psychoactive drugs prescribed to treat hyperactivity and attention deficit disorder (ADHD) by acting on levels of the neurotransmitter dopamine. Modafinil (2-diphenylmethylsulfinyl-acetamide) was developed to treat narcolepsy, and acts on norepinephrine. Prozac is a SSRI developed to treat depression. The benzodiazepine group (including Valium, Librium, Mogadon) was developed to treat anxiety, and act on gamma aminobutyric acid (GABA), the primary inhibitory neurotransmitter of the brain. These psychoactive drugs are used to treat anxiety, depression, various psychopathologies and narcolepsy. However, they may also improve physical and mental performance, reduce exam stress, improve attention and produce a feeling of well-being in "normal" healthy individuals. There is also a growing number of learning and memory enhancing drugs available, such as donepzil or ampakines, that are used both for therapy and for enhancement.

15. It has been estimated that about fifty percent of our traits are pre-encoded by our genes, which are also responsible for the formation of the structures through which we acquire, modify, store and retrieve information. Learning and experience are responsible for generating the remainder of our traits.

16. Kramer (1993; 2002).

17. Diller (2002); Breggin (2000; 2003).

18. These are explored by powerful scanning technologies including Positron Emission Tomography (PET) and functional Magnetic Resonance Imaging (fMRI). See Salleh (2003) and Sowell (2003).

19. Rothman and Rothman (2003).

20. Parens (1998).

21. Caldicott (1998) supports this approach in his report to the Nuffield Council on Bioethics: "A broad and humanistic perspective may be considered to have two basic requirements: respect for human beings and human dignity, and the limitation of harm to, and suffering of, all human beings".

22. Parens (2002). See also Parens (1998a).

23. Kass (2003) has suggested that high self-esteem is earned by the person who has worked hard and consistently but not by those who "cut corners" with pharmacology.

24. Root Wolpe (2002; 2002a; 2003).

25. The importance of ignorance and forgetting has also been stressed by Daniel Dennett (1983).

26. Elliott (2000). See also Elliott (1998).

27. Kass (2003).

28. There is a new utopian movement growing around these drugs; some of these views may be found at the website http://nootropics.com/smartdrugs/brainviagra.html . See also the Paradise Engineering website: http://www.bltc.com/ .

29. The utopian vision promoted by "paradise engineers" is disturbingly close to the dystopian vision of Huxley (1932).

30. It is worth recalling H.L. Mencken's remark "for every complex problem there is a solution that is simple, neat and wrong".

31. Such interventions will pose a different set of more serious ethical questions and ethical challenges but they are in the future, as Pinker (2003) writes "my point is not that genetic enhancement is impossible, just that it is far from inevitable," and with regard to having designer babies, "these traumatic and expensive procedures are not likely to be available soon" ...and... "we can deal with the ethical conundrums if and when they arise".

32. On balance with therapeutic psychoactive drug use good appears to outweigh harm, but their use must continue to be debated by social critics independent of government, professional or industry regulatory agencies and stakeholders. Partly because of the high profitability of this enterprise and partly because of the strength of the connections between drug companies, government and health-professionals, is it not sufficient to have oversight or regulatory powers delegated solely to any of these interest groups. Monitoring should be robust where drugs are used with individuals under the age of eighteen, the elderly or other vulnerable groups or where advertising promotes enhancement.

33. For further argument in support of a precautionary approach to human germ line genetic engineering, see Grey (2005). See also Grey (1996).

CHAPTER FIVE

Cosmology And Theology

John Leslie*

Reasoning known as the cosmological argument (Burrill 1967; Craig 1979, 1980; Hepburn 1967) tries to justify belief in God by pointing to the existence of the cosmos, its causal orderliness, and alleged evidence of its being in some sense designed to include life and intelligence. [Often the appeal to such evidence is instead called the argument from design, or the teleological argument.] Some cosmologists believe, however, that the existence and order of the cosmos can be accounted for scientifically. Its life-permitting character might itself, they consider, be explained through its being divided into multiple domains worth the name of "universes". These could vary randomly in their features, ours being one of the perhaps very rare ones in which life had any chance of evolving. As the anthropic principle reminds us, only the life-permitting universes could give rise to observers. They should hesitate before concluding that an omnipotent, omniscient, all-creating person had made their surroundings life-permitting.

Philosophers, too, have doubted that so remarkable a person would be needed to explain such affairs, or that this person's own existence could be any less in need of explanation. They may here conceive God in a way not everybody would accept, interactions between cosmology and theology often depending on which picture of God is preferred. Such interactions include discussions of the nature of time and of the human mind, and of whether intelligent life is widespread in the cosmos.

Why is there a Cosmos?

People disagree over whether the sheer existence of the cosmos could call for explanation. Some have held that it always would, no matter how long the cosmos had existed; others that it never would; and still others that this would depend on whether the cosmos had existed eternally, or else on the nature of time.

It is nowadays usually believed that our universe came out of a Big Bang, a violent explosion occurring perhaps fifteen billion years ago. Thinking that the Bang could be explained only by God, the atheistic Hoyle (1950) preferred a universe existing eternally in a Steady State: it expanded, but new hydrogen atoms constantly materialized to keep the average cosmic density the same. Although at first offering no explanation for the new atoms, he thought them much less of a difficulty than the materialization of everything at once. Among theists, Pius XII (1952) then agreed that a Bang could indeed be counted as specially strong evidence of God's creative activity, yet many theologians protested that this activity should not be particularly associated with any first moment of a universe's existence. Tables, for instance, would vanish immediately if God failed to "conserve" their existence through exercise of his creative power.

When Hawking (1987, 1988) suggested that his own cosmology left "no place for a creator" since it made "What happened before the Bang, to cause it?" comparable to "What's Earth like to the north of the North Pole?", the theologians renewed their protests. They accompanied them with quotations from Saint Augustine, who had written that God created time and the world together. It is a theological commonplace, though, that God could be described as "creator" even of an eternally existing world. Both in theology and in philosophy, talk of creation or causation does not necessarily assume the temporal priority of creative or causal agencies. Descartes emphasized that in calling God "the cause of himself" he meant only that God's eternal existence was guaranteed by God's nature.

G.Gamow and W.B.Bonnor (both reprinted in Leslie ed. 1990) had meanwhile joined Hoyle in his eagerness to avoid a cosmic beginning. Gamow's article of 1954 favoured a universe which had contracted for infinitely long before rebounding in a Big Bang, whereas Bonnor's of 1960 proposed infinitely many oscillations, the rebounds always occurring before the cosmic material reached infinite densities, which Bonnor called "signs of error". In contrast, Milne (1952) welcomed the Bang as evidence of God's hand. He maintained, too, that the universe at the time of its creation had to be point-like and therefore infinitely dense. Creation of a spatially extended universe was a logical impossibility.

This last contribution to the debate was an outright blunder, there being no contradiction in the idea of creating something extended. Other contributions are harder to evaluate, for intuitions about what should be viewed as a universe's "natural state" -- where this means something not calling for explanation by a divine person or any other external factor -- can be defended or attacked only very controversially. Grünbaum (1990, and in Leslie ed. 1990) thinks it perverse to imagine that tables or the entire universe would disappear if God did nothing to prevent this; in his firm opinion, to discover what happens naturally we must look to see what actually occurs; while Hume (1739) holds that no cause would be needed even for the abrupt entry of a universe into a time which had previously been flowing. Yet one main message which many find in Hume's writings is that we can never learn anything from experience unless helped by various unprovable basic principles: for instance, that the future is likely to resemble the past. Now, one such basic principle might be the need to deny that anything -- let alone an entire universe -- could come to exist for no reason whatever. True enough, quantum theory is often thought to tell us that the universe is indeterministic in some absolute way, but not even this would be a clear declaration that its existence could be utterly reasonless. Again, people can accept quantum laws without thinking that their operation is "natural" in the technical sense of not being a product of God's will.

If we saw a problem in a universe's existence, could it be removed by taking that universe to have existed for infinite past time? Some argue that this would allow its presence at any one instant to be explained by its presence at earlier instants; yet would they then say the same about, e.g., the notion that a particular book had existed eternally? Presumably not. Those Moslems who think the Koran eternal typically consider its existence due to an eternal divine decree. "It exists because it always has done" is not what they think. Moreover, there might be some force in the Kalam Cosmological Argument (Craig 1979) which opposes the world's eternity on the grounds that no infinite series of years could have been traversed to reach the present day. [A complicating factor is that whether the Big Bang's earliest, hottest stages were of finite or infinite duration might depend on one's choice of clock. What if infinitely many events occurred during those stages? The clocks most appropriate to timing the events might then show the stages as taking infinitely long to unfold. As measured by ours, such clocks would tick ever more slowly as the universe cooled.]

A better approach could be that of the quantum cosmologists, Vilenkin (1982) and Hawking (1987) for example, who speculate that quantum theory can give probabilities for worlds of certain kinds to exist. Many recent cosmological models have been inspired by E.P.Tryon's idea of 1973 (Leslie ed. 1990) that even very large universes could begin their existence by taking advantage of quantum indeterminism. Each universe would "cost" little or nothing since the energy tied up in all its particles could be cancelled by their gravitational potential energy, which is standardly treated as *negative energy*. They could therefore be akin to the quantum fluctuations in which individual particles such as electrons exist fleetingly by "borrowing" energies too small to upset quantum theory's rather disorderly balance sheets. It is at present unclear whether this could make sense except against the background of an already existing space, or at least a "space-time foam" lacking clear distinctions between space and time (see Atkins 1992; Craig and Smith 1993; Halliwell 1992; Russell, Murphy and Isham eds. 1996). A common verdict, though, is that even if no such background were needed, there could still be a problem of why quantum laws applied.

Causal Orderliness

Some read Kant as arguing that various principles which we apply to the world, for instance when we view it as causally ordered throughout, are valid only as giving insight into how we necessarily see things. Could this mean that the world was not itself ordered causally, the situation instead being that our unconscious minds were super-geniuses regulating everything we fancied we saw, to ensure that it all obeyed the laws of nature (quantum electrodynamics, general relativity, or whatever)? This would be bizarre, which leads many of Kant's admirers to suspect that he intended something more subtle. Admittedly the laws which people tend to regard as inviolable may some day break down; that is a possibility which cosmologists find easy to defend, familiar as they are with the idea of cosmic "phase transitions" comparable to the change from ice to water and then to steam; but none the less, it remains obvious that the world up to the present moment has had considerable causal orderliness. Now, is this something that calls for explanation?

In the mid twentieth century a widespread opinion was that explaining any one causal law, for instance that heated water changes to steam, could proceed only by appeal to some more basic causal law such as that faster-moving molecules find it easier to break free from one another. The very most basic laws could not possibly be explained, therefore. They would concern mere regularities: as a matter of brute fact, events of one kind would always (or very often) be succeeded by events of some other kind. Discussing the idea that various individuals can know hidden cards by "extra-sensory perception", A.J.Ayer (1970) writes that the only thing which would be remarkable here would be someone's being "consistently rather better at guessing cards than the ordinary run of people"; the fact of doing "better than chance" would prove "nothing at all". If everybody just did know all about playing cards without looking at them, then there would be no mystery in this.

Recently, many philosophers have found such an approach dissatisfying. It is usually ascribed to Hume (1739), yet Hume suggests that causation's patterns would have characterized various events which did not actually occur. As a matter of causal necessity, a window (for example) would have broken had a brick been thrown at it, or would have remained intact had a mere peanut been thrown. This is often thought obviously right, the problem then being how causal patterns could have a *necessity* which was not just a matter of what always in fact occurred.

Might physics and cosmology throw light on any such problem? Quantum theory suggests to many people that we should think in terms of propensities, tendencies, rather than of absolutely firm necessities (a brick might "quantum-tunnel" across a window without smashing it, but this would be exceedingly unlikely) and that the Big Bang grew from a "quantum foam" in which causation as ordinarily conceived could not have acted since there was no firm direction of time. But all this leaves the fundamental issue largely untouched. Why does our world ever obey anything worth the name of a physical law? Or, to express the point differently (Wigner 1960), why "the unreasonable effectiveness of mathematics in the natural sciences"? Why does our world have the kind of elegance which made Jeans (1930) talk of a mathematically minded creator? While no logically possible world could violate mathematical principles, it is easy enough to imagine worlds in which they had little application.

When authors present a divine person or divine creative principle as responsible for the world's existence, they sometimes view causal orderliness as a matter directly attributable to this person or principle (Swinburne 1968 and 1979; Leslie 1979). Whitehead proposed instead (1938) that all things in nature in some sense strive to achieve aims, even atomic particles enjoying some very low level of awareness -- a theme echoed by the physicist D.Bohm (Bohm and Hiley 1993, chapter 15).

Note that the alleged problem of why there are causal laws is rather different from the alleged problem of why anything ever moves or changes, which is what led Aristotle to propose a divine prime mover.

Design and Fine Tuning

Living organisms provide seemingly overwhelming evidence of divine design, their parts forming immensely complicated mechanisms which permit them to survive and to reproduce themselves. To say "That's how they just happen to be" would be silly, although Philo in Hume's *Dialogues* (1779), sometimes thought to speak for Hume himself, may at times be guilty of saying it. Had Darwin said it, then he would never have discovered his famous way of undermining the seemingly overwhelming evidence: his theory of evolution, that is to say. Darwin's theory is now known to be right, so supporters of design seek their evidence elsewhere. They point to how the world's laws combine to make Darwinian evolution possible (Henderson 1913 supplies an early example of this). Why, they ask, is there a friendly environment in which extremely complex living machinery could appear after long ages, through selective inheritance of more and more complicated genes?

They can here direct our attention not just to the general fact of causal orderliness as discussed above but also (Leslie 1989, chapter 3) to the fortunate effects of particular causal principles. Consider, for instance, the principles met with in quantum theory. As well as allowing apparently dissipated wave-energy to concentrate itself so that it can do useful work, quantum laws ensure that atoms come in standardized types, making the genetic code possible. Similar things can next be said of the laws of special relativity, and of various laws controlling elementary particles.

However, much more attention has been directed towards the apparent "fine tuning" of fundamental cosmic parameters: the strengths of physical forces, the masses of elementary particles, the expansion speed and degree of turbulence at early moments in the Big Bang, and so forth (Barrow and Tipler 1986; Davies 1982; Ellis 1993; Leslie

1989, chapter 2; Leslie ed. 1990; Polkinghorne 1986; Rolston 1987). For example, it appears that electromagnetism, gravity, and the two main forces which control the atomic nucleus, had all of them to have strengths which fell inside very narrow limits if there were to be any stars of the long-living, steadily burning sort: the sort which encourage life to evolve. Again, life's complex chemistry appears possible only thanks to very precise adjustment of the masses of the neutron, the proton and the electron.

For there to be life of any readily imaginable kind, anything up to several dozen factors can appear to have needed fine tuning. Because the number of factors to be listed seems so large, this supposed evidence of design can survive many doubts about what exactly should be on the list. Take the case of the early cosmic expansion speed. It is often held that cosmic inflation, a brief burst of tremendously rapid expansion occurring early in the Bang, resulted in a universe whose subsequent more leisurely expansion was in no need of tuning. Yes, the expansion speed after inflation had to fall inside very narrow limits for stars to be able to form but, it is often said, inflation more or less forced the speed to fall inside those limits. To this we might reply that inflation itself stood in need of very delicate tuning, yet we could instead simply drop the expansion speed from the list, pointing out that plenty of other items remained on it. [We could find the list impressive without claiming knowledge of all logically possible universes, most of them presumably with properties very distant from those of our universe, and of what proportion of these possible universes were life-permitting. Imagine that a bullet hits a fly surrounded by a large empty area. The bullet's trajectory needed fine tuning to achieve this result, which can help to show that a marksman was at work. It can help to show it regardless of whether distant areas are all of them so covered with flies that any bullet striking them would hit one. The crucial point is that *the local area* contained just the one fly.]

It is sometimes held that we could avoid belief in fine tuning by believing instead in various exotic life-forms (Feinberg and Shapiro 1980). Rather than being based on chemistry (which means, in effect, on electromagnetism), intelligent organisms might be based on the strong nuclear force so that they could inhabit neutron stars. Alternatively, they might be plasma beings inside the sun, or complex patterns in frozen hydrogen, or intricately organised interstellar gas clouds. None of these intelligent life-forms would dream of arguing that chemistry, something possible only through fine tuning, was necessary to intelligent life. Yet people willing to believe in such strange life-forms could still say that much tuning was required for there to be neutron stars, suns, planets covered with frozen hydrogen or interstellar clouds. It is often reasoned (see Rozental 1988 in particular) that a universe taken at random from among the apparent physical possibilities would almost certainly lack such objects. It would be likely either to collapse in a Big Crunch after a very brief, intensely hot career, or else to expand so fast that any matter which it contained soon became too rarefied to form clouds, let alone stars and planets. It could well consist almost entirely of light rays or black holes.

Various other doubts about fine tuning and the need to explain it are fairly easily dismissed. For instance, it seems wrong to reason (i) that no possible evidence of design could have any force *because we can see only the one universe*, and therefore cannot know whether its patterns are at all extraordinary, or (ii) that all possible patterns *would be equally probable*, just like all possible hands of cards, or (iii) that *probabilities depend on repetitions being possible* whereas the universe is unrepeatable: a universe can occur only once. Such reasoning delivers the strange conclusion that not even the words "God designed all this", written on every rabbit, tree and snowflake, could be in special need of explanation. It forgets such facts as that a hand of cards which includes four queens, four kings and four aces can, thanks to the possibility of cheating, be considerably more probable than many others. And the claim "that a universe can occur only once" itself runs into trouble: see the next section, "Multiple Universes".

Again, it would surely be wrong to protest that fine tuning needs no explanation "since if the universe hadn't been tuned in appropriate ways, then there'd not have been anybody to consider the affair". What would you think of the man who, untouched by all the bullets of a fifty-marksman firing squad, failed to suspect that the marksmen had wanted to miss him, commenting instead "that he'd otherwise not be alive to discuss anything"?

A divine designer's influence might be limited to creating a universe with life-permitting laws and fine-tuned force strengths, particle masses, etcetera. Some (e.g., Ward 1996b) argue, however, that God could be expected to influence the course taken by Darwinian evolution, ensuring that various crucial events occurred in favourable ways. [Assuming that quantum physics makes the reign of natural law into something only probabilistic, not deterministic, then there is actually some difficulty in deciding whether God, by ensuring that such and such an event occurred in the most favourable of various ways which quantum physics allowed, would be "intervening miraculously".] Also, it is sometimes thought that God created a universe in which the evolution of intelligent, truly conscious minds, and the workings of those minds during free decision-making, are at least in part inexplicable by physical laws (Swinburne 1986).

Multiple Universes

If by "universe" you mean Absolutely Everything, then there must be just a single universe. However, people often picture the cosmos as containing numerous huge domains, very varied in their characters and largely or entirely isolated from one another. Now, "universes" is what they typically call them nowadays. Understood in this way, universes can be used to explain any observed fine tuning without introducing a divine designer. While most universes could well be hostile to intelligent life, observers would clearly have to find themselves in the life-permitting ones.

Numerous universe-generating mechanisms have been proposed (Atkins 1992; Barrow and Tipler 1986; Barrow 1988; Halliwell 1992; Leslie 1989; Leslie ed. 1990; Linde 1990; Rees 1997; Rozental 1988; Smolin 1997). Universes could be successive cycles of an oscillating cosmos (Big Bang, Big Crunch, Big Bang, etcetera). They could be huge areas of a gigantic, perhaps infinite cosmos. They could be the "worlds" of Many-Worlds Quantum Theory, which says that reality continually branches, every alternative allowed by quantum laws occurring in some branch or other. They could be quantum fluctuations in a pre-existing space or in a space-time foam. Or they might "quantum-tunnel from nothing" (if that makes sense), or bud off from other universes, or form bubbles in which expansion speeds had slowed inside a cosmos which was perpetually inflating. They could even be born in the depths of black holes, then expanding into spaces of their own without disturbing their parent universes.

All this could help an opponent of divine design only if the universes differed widely so that sooner or later, somewhere, one or several of them might be expected to be fine tuned in life-permitting ways. Difference-generating mechanisms are easily invented, however. An early suggestion was that an oscillating cosmos would "forget" its properties in the quantum-fuzzy depths of its Big Crunches. Among many later suggestions, perhaps the most plausible is that tiny domains form in the cosmos like ice crystals on a pond, each domain then being enormously enlarged (to "universe size") by cosmic inflation so that living beings deep inside it cannot see the other domains. Wide differences between the domains can readily be attributed to *scalar fields*. Having no directionality such as makes a magnetic field obvious to a compass needle, scalar fields are hard to detect, yet it is now standardly considered that such fields tore apart the initially unified forces of nature (electromagnetism and the nuclear forces, for instance) and gave them their various strengths, also causing various types of particle to become massive to differing degrees. Appearing early in the Big Bang, the scalar fields could have differed randomly from place to place because different field intensities were more or less equal in their potential energies (which are what physical systems try to minimize, like balls rolling down into valleys).

Belief in multiple domains, alias universes, and in the likelihood that they differ widely, is nowadays extremely common among cosmologists. It is considered quaint to assume that all of reality must be like the region visible to human telescopes. This would have pleased Hume's Philo, who cautioned us against any such assumption (Hume 1779). A reality consisting of infinitely many, very varied universes may actually be thought simpler than the alternatives. Why, after all, should a universe-generating mechanism operate only a limited number of times? And if it operated again and again, why should it produce identical results on each occasion?

This does not mean that belief in divine design must be abandoned. While the manner in which our universe appears "fine tuned for permitting life to evolve" could encourage a story about many widely differing universes, it could equally well support belief in a designer. The fact remains, though, that the designer does not supply the sole plausible explanation for any fine tuning. This is largely because universes, like ice crystals, could differ widely while remaining identical in the fundamental laws they obeyed.

The Anthropic Principle

In the early 1970s, Brandon Carter stated what he called "the anthropic principle": that what we can expect to observe "must be restricted by the conditions necessary for our presence as observers" (Leslie ed. 1990). Carter's word "anthropic" was intended as applying to *intelligent beings in general.* The "weak" version of his principle covered the spatiotemporal districts in which observers found themselves, while its "strong" version covered their universes, but the distinction between spatiotemporal districts and universes, and hence between the weak principle and the strong, *could not always be made firmly*: one writer's "universe" could sometimes be another's "gigantic district". Moreover, the necessity involved was never -- not even in the case of the "strong anthropic principle" -- a matter of saying that some factor, for instance God, had made our universe *utterly fated* to be intelligent-life-permitting, let alone intelligent-life-*containing*. However, all these points have often been misunderstood and, at least when it comes to stating what words mean, errors regularly repeated can cease to be errors. Has Carter therefore lost all right to determine what "anthropic principle" and "strong anthropic principle" really mean? No, he has not, for his suggestion that observership's prerequisites *might set up observational selection effects* is of such importance. Remember, it could throw light on any observed fine tuning without introducing God. Everything is thrust into confusion when people say that belief in God "is supported by the anthropic principle", meaning simply that they believe in fine tuning and think God can explain it. As enunciated by Carter, the anthropic principle does not so much as mention fine tuning.

Being aware of possible "anthropic" observational selection effects can encourage one set of expectations, and belief in God another set. If suspecting that Carter's anthropic principle has practical importance, you will be readier to believe (i) that there exist multiple universes and (ii) that their characteristics have been settled randomly, some mechanism such as cosmic inflation ensuring that all was settled in the same fashion throughout the region visible to our telescopes. True, the believer in God can accept these things too, yet he or she may feel far less pressure to accept them. Even if there existed only a single universe, God could have fine tuned it in ways that encouraged intelligent life to evolve.

A possible argument for preferring the God hypothesis runs as follows. A physical force strength or elementary particle mass can often seem to have required tuning to such and such a numerical value, plus or minus very little, *for several different reasons*. Random variations from universe to universe might explain why it took any particular value somewhere or other, yet how could they account for the fact that one and the same value satisfied many different requirements? Why is such consistency possible? Why does electromagnetism, for example, not need to have one strength to allow atoms to be stable, and another strength for stars to burn at a life-encouraging rate, and yet another to permit carbon (quite probably crucial to life) to be produced plentifully? Here a religiously minded physicist could think in terms of many possible fundamental theories, God selecting a theory which permitted life's requirements to be fulfilled without contradictions.

God

It is sometimes protested that God cannot adequately explain the existence and orderliness of the cosmos, for the following reason: that God's own existence, and the orderliness which would have to characterize his mind before he could bring order to anything else, would in turn need explanation. How might theists reply?

It might seem that an infinitely knowledgeable divine mind would be infinitely harder to explain than any finite cosmos or than one which, while infinite in both time and space, was still limited in, for example, the number of its dimensions -- whereas the divine mind would know every possible universe, including those with a million billion dimensions. Again, such a mind might be thought to go infinitely far beyond any evidence we could collect. [If you saw a pound of butter rising on a balance pan, would you conclude that it was being outweighed by an infinitely heavy weight?] It can be replied, however, that an all-knowing mind *would be in a way extremely simple*, as can be seen by how a single word -- "everything" -- can describe what it knows; and such an all-knowing mind, it could next be said, might well be expected to create a complex cosmos for the sake of all the living beings in it. But while the combination of the cosmos and a divine creator might be considered simple for reasons such as these, there remains the difficulty that a complete and utter blank could well be thought simpler still. Note that the "ontological argument" which tries to prove God's existence from his mere notion is generally dismissed today. There would seem to be no contradiction in the idea that a perfect being *was a logical possibility only,* not an actual existent.

On the other hand, logical possibilities can be *real* without being *actual existents,* and once this is appreciated their reality can be seen to be guaranteed. If God or anything else is a logical possibility, then that is an unconditional fact. It is eternally the case, non-fictitious, genuine, real, that this or that is indeed a logical possibility. [How odd it would be to fancy that thinkers, for instance, could become *really logically possible* only after they had come into existence and developed logics! Before thinkers evolved, the sun which helped them to evolve was logically possible, surely. It was not like a round square, and nor were the thinkers.] Furthermore, some logical possibilities are such that their failure to be actualized -- their absence from the realm of actually existing things -- can be thought needed, ethically required, in an eternal and unconditional way. Take the case of a logically possible world consisting solely of people in torment. To declare that the actual existence of such a world would be evil is the same as calling its non-existence needed or ethically required, and it could seem utterly wrong to add "just so long as there exists somebody who is contemplating the affair, somebody with a moral duty to prevent the existence of such a world". Likewise, to say that a world full of interest and happiness would be a good thing is to say that the existence of such a world is ethically required, and would (presumably) be so regardless of whether anyone ever contemplated that fact. Might we point to these matters when trying to account for the existence of God or of the cosmos?

Certainly, the concept of an ethical requirement is distinct from that of a requirement which is fulfilled; yet cannot a flower be red, despite how the concept of redness differs from that of being a flower? Flowers are "the right sort of reality" for being red. They are "in the right ball park". The idea of a red flower is not conceptually confused. And rather similarly, it can be argued, a requirement for the existence of something, for instance a divine mind, might be the right sort of reality to carry responsibility for this something's actual existence, *even if it were an ethical requirement*. There is no conceptual confusion here. So long as it was recognized that no logical necessity was involved, it could actually be suspected that some ethical requirement (or consistent set of requirements) carried such responsibility *necessarily*. Compare, perhaps, the necessity that red as we experience it (say, in an after-image produced by a bright light) is nearer to purple than to blue. This can be argued to be an absolute necessity, without being a logical necessity. Again, it can be argued that various states of mind are necessarily in themselves worth having, without this being logically demonstrable.

An ethical requirement is, at any rate, what Ewing (1973, chapter 7) proposes as the ground of a divine person's reality. As a matter of necessity -- necessity which is absolute despite not being provable by logicians -- the unconditionally real ethical need for a divine mind to exist is adequate, Ewing suggests, to ensure its eternal existence. And a very similar theme is central to the long neoplatonic tradition in theology. This treats "God" as the name not of any mind, but of the supposed fact that the world owes its existence to its ethical requiredness (Leslie 1979, 1989 chapter 8; Mackie 1982, chapter 13; Levine ed. 1997; Tillich 1953-63).

Ewing's picture may at times be hard to separate from that of the neoplatonists. For suppose we joined the pantheist Spinoza in thinking (to take an interpretation of his writings which can seem to make sense of them) that all the complexities of the cosmos are simply the complex thoughts of a divine mind, so that your consciousness and mine, or the consciousness of a bird or of a bat, is just the divine knowledge of what it is like -- exactly what it feels like -- to be particular living beings with strictly limited power and knowledge. Suppose we also adopted the belief (which can again be the suggested by Spinoza's writings) that the divine mind exists *because this is ethically required*, rather than through any logical necessity which has nothing to do with good or bad. There might then be no real difference, so far as concerned the situation in which we believed, between our declaring (a) that God was a divine mind, as Ewing thought, or (b) that God was the cosmos, as Spinoza thought, or (c) that, as neoplatonists think, God is a creative force or principle: the principle that a supreme ethical requirement (or consistent set of requirements) is responsible for the existence of the cosmos.

Sure enough, this disregards distinctions which have appeared important to many people. For many variants on pantheism and on neoplatonism, consult Forrest (1996),

Laird (1940), Levine (1994) and Whitehead (1938). Sometimes Spinoza's eternal and all-knowing divine mind is replaced by one which constantly improves its power and its knowledge. Sometimes God is thought to some degree separate from the cosmos, instead of just being the cosmos or the creative ethical requiredness of the cosmos. Sometimes the notion that the natural world is alive, and that it strives after value, plays a greater role than the idea that value is actually achieved. Sometimes identification of the cosmos with God is oddly taken to imply the illusoriness of most of the things we think we see.

Time and the Human Mind

Believing (as Spinoza appears to) that we are all parts of a divine mind, we might think we could answer the theological problem of evil: the problem of why the cosmos contains so many items which can seem so very unsatisfactory. Knowing everything, or (if this is different) everything in the least worth knowing, a divine mind could know many things which might be very little worth knowing if they were taken in isolation. It might know, for example, not only all the detailed structure of innumerable universes, but also exactly how it would feel to be each of the intelligent living beings in those universes. Knowing this eternally, it could none the less know what it felt like to be engaged in actual struggles, in constant ignorance of what the next moment would bring: see Williams (1951) to gain further insight into the theory about time's flow which would be involved here, a theory often adopted because it appears to reflect Einstein's theory of relativity. While such items of knowledge might be nowhere near the best that it possessed, the divine consciousness could still be better for not being ignorant of them.

A competing approach to the problem of evil is often preferred, though. Instead of viewing themselves as elements of a divine mind, many religious people think they exist separately from that mind and are given absolutely free choice of whether to join it in a heavenly hereafter.

Absolutely free choice, as they conceive it, depends on time's flowing in a way which Einstein rejected when he wrote of the world as having "a four-dimensional existence". [Suppose that you are about to choose "with absolute freedom" whether to give up smoking. It cannot already be true, they say, that "at points a little further along the fourth dimension" you are lighting a cigarette.] It probably also demands a fairly strong division between the operations of human minds and those of material objects.

Extraterrestrials

Belief in God need not involve believing that our universe is crammed with intelligent life from side to side and from start to finish, or that God will save humans from driving themselves to extinction through polluting their planet or by germ warfare, or perhaps (Rees 1997, chapter 12) by experiments at extremely high energies. The fraction of our universe which we can see contains many hundred million trillion sun-like stars. Even if only a small proportion of these had hospitable planets, intelligent beings could well exist in tremendously many places. What is more, the universe as a whole might be very much larger, perhaps infinitely larger, and there may be up to infinitely many other universes. Wishing for intelligent beings to exist in large numbers, a divine person could have them without ensuring that humans survived long enough to colonize their entire galaxy.

We have failed to detect extraterrestrials, although calculations suggest that an intelligent species could spread across its galaxy in a few million years. This, together with our observed position in the midst of a population explosion, might reinforce whatever other grounds we had for thinking that rapid extinction, not galactic colonization, was the likely fate of an intelligent species. Suppose human extinction occurred during the next century. Roughly ten per cent of all humans who had ever been born would be alive when it occurred. On the other hand, if humans spread right across their galaxy then perhaps well under a thousandth of one per cent would have lived during that period. This may be found disturbing. [The point is that one ought to hesitate before adopting theories whose truth would have made one's own observations highly unlikely, when other theories would have made them fairly likely. It is a point first noted by Brandon Carter; for a discussion drawing very pessimistic conclusions from it, see Gott (1993). Various reasons in favour of guarded optimism are given by Leslie (1996), reasons centred on the fact that our universe is probably indeterministic, so that the number of humans who will ever have lived is something which has not yet been fixed.]

For religious people asking whether humans will soon be extinct, here is a point to bear in mind. If you hope to solve the theological problem of evil, then you have to assume that there are strong reasons *against* divine intervention to prevent calamities.

Bibliography

- Atkins, P.W. 1992. *Creation Revisited,* Oxford: W.H. Freeman.
- Ayer, A.J. 1970. " Chance," chapter 7 of *Metaphysics and Common Sense*, Freeman, Cooper: San Francisco.
- Balashov, Y.V. 1991. "Resource Letter AP-1: The anthropic principle," *American Journal of Physics,* 59: 1069-1076.
- Barrow, J.D., and Tipler, F.J. 1986. *The Anthropic Cosmological Principle*, Oxford: Clarendon Press.

- Barrow, J.D. 1988. *The World within the World,* Oxford: Clarendon Press.
- Bertola, F., and Curi, U., eds. 1993. *The Anthropic Principle,* Cambridge: Cambridge University Press.
- Bohm, D., and Hiley, B.J. 1993. *The Undivided Universe,* London: Routledge.
- Burrill, D.R. 1967. *The Cosmological Arguments,* New York: Doubleday.
- Carr, B.J., and Rees, M.J. 1979. "The anthropic principle and the structure of the physical world," *Nature,* 278: 605-612.
- Carter, B. 1989. "The anthropic principle: self-selection as an adjunct to natural selection," 185-206 of S.K.Biswas et al., eds., *Cosmic Perspectives,* Cambridge: Cambridge University Press.
- Craig, W.L. 1979. *The Kalam Cosmological Argument,* London: Macmillan.
- Craig, W.L. 1980. *The Cosmological Argument from Plato to Leibniz,* London: Macmillan.
- Craig, W.L., and Smith, Q. 1993. *Theism, Atheism and Big Bang Cosmology,* Oxford: Clarendon Press.
- Davies, P.C.W. 1982. *The Accidental Universe,* Cambridge: Cambridge University Press.
- Davies, P.C.W. 1992. *The Mind of God,* New York: Simon and Schuster.
- Ellis, G. 1993. *Before the Beginning,* London: Bowerdean Press/ Marion Boyers.
- Ewing, A.C. 1973. *Value and Reality,* London: Allen and Unwin.
- Feinberg, G., and Shapiro, R. 1980. *Life Beyond Earth,* New York: William Morrow.
- Flew, A. 1966. *God and Philosophy,* London: Hutchinson.
- Forrest, P. 1996. *God without the Supernatural,* Ithaca: Cornell University Press.

- Gott, J.R. 1993. "Implications of the Copernican principle for our future prospects," *Nature*, 363: 315-319.
- Grünbaum, A. 1990. "Pseudo-creation of the big bang," *Nature,* 344: 821-822.
- Halliwell, J.J. 1992. *Quantum Cosmology,* Cambridge: Cambridge University Press.
- Hassing, R.F., ed. 1997. *Final Causality in Nature and Human Affairs,* Washington: The Catholic University of America Press.
- Hawking, S.W. 1987. "Quantum cosmology," pages 631-651 of S.W.Hawking and W.Israel, eds., *Three Hundred Years of Gravitation,* Cambridge: Cambridge University Press.
- Hawking, S.W. 1988. *A Brief History of Time*, New York: Bantam Books.
- Henderson, L.J. 1913. *The Fitness of the Environment,* New York: Macmillan.
- Hepburn, R.W. 1967. "Cosmological Argument for the Existence of God," 234-237 of P.Edwards, ed., *The Encyclopedia of Philosophy, Vol. 2,* New York: Macmillan.
- Hetheringtor, N.S., ed. 1993. *Encyclopedia of Cosmology,* New York: Garland.
- Hoyle, F. 1950. *The Nature of the Universe,* Oxford: Blackwell.
- Hume, D. 1739. *A Treatise of Human Nature*, London.
- Hume, D. 1779. *Dialogues concerning Natural Religion,* London.
- Jeans, J. 1930. *The Mysterious Universe*, London: Macmillan.
- Kragh, H. 1996. *Cosmology and Controversy,* Princeton University Press: Princeton.
- Kuiper, B.H., and Brin, G.D. 1989. "Resource Letter ETC-1: Extraterrestrial Civilization," *American Journal of Physics,* 57: 12-18.

- Laird, J. 1940. *Theism and Cosmology,* London: Allen and Unwin.
- Leslie, J. 1979. *Value and Existence*, Oxford: Blackwell.
- Leslie, J. 1989. *Universes*, London and New York: Routledge.
- Leslie, J., ed. 1990. *Physical Cosmology and Philosophy*, New York: Macmillan.
- Leslie, J. 1996. *The End of the World: the science and ethics of human extinction,* London and New York: Routledge.
- Levine, M. 1994. *Pantheism,* London and New York: Routledge.
- Levine, M., ed. 1997. *Pantheism* (special issue, volume 80, of *The Monist*).
- Linde, A.D. 1990. *Inflation and Quantum Cosmology*, San Diego: Academic Press.
- Mackie, J.L. 1982. *The Miracle of Theism*, Oxford: Oxford University Press.
- Margenau, H., and Varghese, R.A., eds. 1995. *Cosmos, Bios, Theos*, La Salle: Open Court.
- Matthews, C.N., and Varghese, R.A., eds. 1995. *Cosmic Beginnings and Human Ends,* Chicago: Open Court.
- McMullin, E., ed. 1985. *Evolution and Creation,* Notre Dame: University of Notre Dame Press.
- Milne, E.A. 1952. *Modern Cosmology and the Christian Idea of God,* Oxford: Clarendon Press.
- Munitz, M.K., ed. 1957. *Theories of the Universe,* New York: Macmillan.
- Munitz, M.K. 1986. *Cosmic Understanding,* Princeton: Princeton University Press.
- Peacocke, A.R. 1979. *Creation and the World of Science*, Oxford: Clarendon Press.
- Parfit, D. 1992. "The puzzle of reality," *Times Literary Supplement,* July 3: 3-5.

- Pius XII (Pope). 1952. "Science and the catholic church," *Bulletin of Atomic Scientists* 8: 142-146, 165.
- Polkinghorne, J. 1986. *One World: the Interaction of Science and Theology*, Princeton: Princeton University Press.
- Rees, M. 1997. *Before the Beginning: Our Universe and Others*, Reading, Mass.: Addison-Wesley.
- Rescher, N. 1984. *The Riddle of Existence,* Lanham: University Press of America.
- Rolston, H. 1987. *Science and Religion*, New York: Random House.
- Rozental, I.L. 1988. *Big Bang, Big Bounce,* Berlin: Springer-Verlag.
- Russell, J.R., Murphy, N., and Isham, C.J., eds. 1996. *Quantum Cosmology and the Laws of Nature,* Notre Dame: University of Notre Dame Press.
- Russell, R., Stoeger, W., and Coyne, G., eds. 1988. *Physics, Philosophy and Theology*, Notre Dame: University of Notre Dame Press.
- Smart, J.J.C. 1989. *Our Place in the Universe*, Oxford: Blackwell.
- Smart, J.J.C., and Haldane, J.J. 1996. *Atheism and Theism*, Oxford: Blackwell.
- Smolin, L. 1997. *The Life of the Cosmos,* New York: Oxford University Press.
- Swinburne, R. 1968. "The Argument from Design," *Philosophy*, 43: 199-211.
- Swinburne, R. 1979. *The Existence of God*, Oxford: Clarendon Press.
- Swinburne, R. 1986. *The Evolution of the Soul*, Oxford: Clarendon Press.
- Tillich, P. 1953-63. *Systematic Theology,* 3 vols., London: Nisbet.
- Vilenkin, A. 1982. "Creation of Universes from Nothing," *Physics Letters,* 117 B: 25-28.

- Ward, K. 1996a. *Religion and Creation*, Oxford: Clarendon Press.
- Ward, K. 1996b. *God, Chance and Necessity*, Oxford: Oneworld Publications.
- Whitehead, A.N. 1938. *Modes of Thought,* New York: Macmillan.
- Wigner, E.P. 1960. "The unreasonable effectiveness of mathematics in the natural sciences," *Communications in Pure and Applied Mathematics*, 13: 1-14.
- Williams, D.C. 1951. "The Myth of Passage," *Journal of Philosophy,* 48: 457-472

---

* Reprinted by permission from:

Leslie, John, "Cosmology and Theology", *The Stanford Encyclopedia of Philosophy (Summer 2003 Edition)*, Edward N. Zalta (ed.), URL = <http://plato.stanford.edu/archives/sum2003/entries/cosmology-theology/>.

AFTERWORD

In the years since the above article first appeared, little has changed in how cosmology relates to theology. In cosmology the main news is as follows. Studies of distant quasars seem to show that the cosmic expansion, instead of continuing to slow down, has recently accelerated. It could well mean that ours isn't an eternal, cyclical world in which Big Bangs are for ever followed by Big Squeezes which "bounce" into new Big Bangs. Nevertheless, this remains very unsure. The acceleration might be an illusion, or at any rate insufficient to prevent an eventual Squeeze. The "dark energy" said to be responsible for the speeded expansion might actually change, sooner or later, to pulling the universe together. In any case, any theological need to explain why there exists a cosmos, not a blank, doesn't depend on denying that the cosmos has existed for ever. A cosmos of infinitely many Bangs and Squeezes could be interestingly different from a blank.

The years have, however, seen more and more cosmologists declaring that we detect but a tiny fragment of Reality. The idea that there exist multiple regions which are sufficiently large, sufficiently separate and sufficiently diverse to be worth calling "separate universes" has become a scientific commonplace. Having successive Bang-Squeeze cycles, each "a new universe" with new properties, is just one among numerous means of putting flesh on this idea. String theory, for example, offers a cosmos split into immensely many regions whose dimensions have become compactified in different ways. These "separate universes" could have properties so varied that the presence of life-permitting properties *somewhere* might be no mystery, regardless of how much "fine tuning" properties need in order to be life-permitting. Since we couldn't find ourselves in a region whose properties *weren't* life-permitting, God as Fine Tuner might be out of a job—as is suggested by the founder of string theory in a recent book (Susskind 2005).

Theologians could, though, try to give God credit for creating a cosmos in which appropriately tuned regions were bound to occur. And they could add that neither string theory nor any other physical theory can answer why there's a cosmos and not a blank.

It does seem that the question needs an answer. Also that the sole possible answer is given by a Platonic creation story: a story saying that the cosmos, the sum total of all existence, is something worthwhile, and that its worthwhileness sufficiently explains its reality. The ethical requirement that there exist a worthwhile scheme of things isn't put into effect, given creative importance, by a benevolent divine being whose own existence needs no explanation. Instead the requirement is itself fully responsible for the reality of the entire cosmos. The patterns of events in our universe are among endlessly much that would be contemplated by a mind that was a reality of the best possible type, an infinite mind that contemplated all that was worth contemplating. Well, it could be that those event-patterns, maybe joined by the event-patterns of innumerable other universes, are nothing but patterns contemplated by such a mind. No possible scientific discoveries could disprove this. And why on earth imagine like your typical theologian that an infinite mind, supposedly omnipotent, has created *outside the patterns of its own thought* a scheme of things containing vastly many other minds, all of them infinitely inferior to itself? What a disappointing thing for omnipotence to do! A Platonic creation story can describe a better situation (Leslie 2001). The Real consists of infinitely many infinite minds.

Each exists because its existence is ethically required. Each creates universe-patterns inside itself when it thinks all the thoughts that are worth thinking.

That such a story can be pleasant is of course no proof of its correctness. Still, neither is it an argument against it. A Platonic creation story which failed to provide a pleasant world-picture—a story which, for example, took fright at the notion of minds that contemplated all that was worth contemplating, or which firmly denied immortality of the sort (or, see Leslie 2007, *sorts*) that you and I could have through existing inside a mind like that—might be altogether too odd. Almost as odd, perhaps, as a Platonic creation story which said that ethical factors had created a horrible cosmos.

None of this proposes that evils are always illusory. It doesn't propose, even, that our universe is secure against large-scale disasters. It could well be that the cosmic expansion will eventually accelerate to a Big Rip which tears space to pieces. Or that humans of the fairly near future, experimenting at ever higher energies, will upset a space-filling scalar field which isn't fully stable. That's one of many nasty scenarios taken seriously by Martin Rees (2003). The resulting bubble of new-strength scalar field would expand at virtually the speed of light, destroying everything. After roughly a hundred thousand years, no more galaxy; after somewhat longer, no more Local Supercluster; etcetera.

References

Leslie, J. (2001) *Infinite Minds*. Oxford and New York: Oxford University Press/Clarendon Press.

---- (2007) *Immortality Defended*. Oxford, Malden Mass., and Victoria, Australia: Blackwell.

Rees, M.J. (2003) *Our Final Century.* London: Heinemann.

Susskind, L. (2005) *The Cosmic Landscape: String Theory and the Illusion of Intelligent Design.* New York: Little Brown.

Chapter Summary

Our universe, as well as having causal orderliness, is fine tuned in life-permitting ways. One and the same parameter, for instance a force strength or a particle mass, often satisfies many different requirements. This could best be explained by the Platonic theory that the cosmos, the sum total of all existence, exists simply because that is ethically required. It could well consist of infinitely many eternal minds, each contemplating the structures of many different universes. Existing among the things contemplated, humans might have immortality. But the human race could soon be extinct, perhaps because of high-energy experiments.

CHAPTER SIX

Positive Logicality:
The Development Of Normative Reason

J. R. Lucas

§1   Non-contradiction

One can think wrong. The fact that after much thought one has reached a conclusion is no guarantee that the conclusion reached is right. Only a very opinionated man would refuse to concede the possibility of error, and once the admission of fallibility is made, the problem of justifying one's beliefs becomes acute. So we formulate our reasons as best we can. But even when formulated, they may fail to convince. Only if people are willing to be reasonable, can they be reasoned with. None so obdurate as those who will not listen to reason, and with them at least it is better to save one's breath than to attempt to convince them. You just cannot argue with people who will not be argued with. With them we can only let them go their way, as did Socrates; *ea chairein.*

And yet. Sometimes even the most unreasonable sophist can be caught out, and made to admit the force of an argument. Socrates was able to make Thrasymachus blush, and Aristotle gave some general rules for arguments which are absolutely incontrovertible. If I admit that All men are mortal and that Socrates is a man, I must then concede that Socrates is mortal. Why? What happens if I do not? If I do not concede that Socrates is mortal but say instead that he is not mortal, having already said that All men are mortal and that Socrates is a man, then I am contradicting myself.

The traditional syllogism

> All men are mortal
> Socrates is a man
> *therefore* Socrates is mortal

is valid [1] because the conjunction of the three propositions

> All men are mortal
> Socrates is a man
> Socrates is not mortal

is inconsistent. A man who utters this inconsistent triad of propositions is guilty of a self-contradiction. Having affirmed any two, he must not go on to affirm the third on pain of inconsistency, but must contrariwise concede the negation of that third proposition. If he affirms the first two he must concede the negation of the third, as in the familiar syllogism. If he had affirmed the second and the third, he would have to deny the first, as in the syllogism

> Socrates is a man
> Socrates is not mortal
> *therefore* Not all men are mortal,

while if he had affirmed the first and the third, he would have to deny the second, as in the syllogism

All men are mortal
Socrates is not mortal
therefore Socrates is not a man.

Simple informal deductive arguments can be defined in terms of inconsistency in much the same way as analytic propositions can be. A proposition is analytic if its negation is inconsistent, and similarly an argument is deductive if it would be inconsistent to affirm the conjunction of the premises and the negation of the conclusion; or, more colloquially, if a man would be contradicting himself if he affirmed the premises and denied the conclusion. We are bound to admit the force of deductive arguments because we should be contradicting ourselves if we did not. Having said that All men are mortal and that Socrates is a man, we cannot refuse to allow that Socrates is mortal, any more than we can resist the claim that All red things are coloured, that All uncles are brothers, that All bachelors are unmarried, or that Either it is raining or it is not. Deductive arguments, like analytic propositions, flow from the Law of Non-contradiction, and thence obtain their incontrovertible validity.

But why should I not contradict myself? It is a free country, and it would be uncivilised to make me bridle my tongue out of deference to Plato or Aristotle. And indeed there are no legal penalties for inconsistency. The sanction is quite another one, namely that if I contradict myself I make myself unintelligible. A speaker must be consistent, or communication breaks down. In most systems of formal logic [2] it is easily shown that if both p and not-p are given, we can prove any other proposition we like; and one definition of the consistency -- the "absolute consistency" of a system -- is that not every proposition can be proved in it.

The same thought is expressed in the colloquial rejoinder "If you would say that, you would say anything". And this is a rebuke, because if a person is prepared to say anything, then anything he says is no better than anything else. Only if there are some things he is not prepared to say does the fact of his saying some other thing signify. Where everything is free, nothing is of value. Propositions acquire meaning in as much as they have scarcity value. Only if Thrasymachus is not prepared to say absolutely anything, will people attend to what he actually does say. Else his words cease to have scarcity value or significance, and cease to be words at all, and become just babble.

Simple informal deductive arguments, therefore, are valid. Anybody who refuses to accept a deductive argument, puts himself out of court by making himself unintelligible. If he wishes to communicate, he must use a language, and abide by the rules of that language, which alone constitute it as a language and not a sequence of meaningless noises. They are therefore utterly incontrovertible, since only if he accepts them can he make himself understood. Even a sophist, even Thrasymachus, could not resist deductive arguments. They are absolutely cogent, and make no demands on a person's being reasonable, but only on his being able to use language. In that sense, man is a *logistikon zoon*, a talking animal, even more ineluctably than that he is rational. Not all men are reasonable, but all who argue with us are of necessity language-users.

Although informal deductive arguments are valid, there may be in some cases a problem of recognising them for what they are. The informal deductive arguments which Plato has Socrates use against his opponents obtain their force from the rules of correct English (or Greek) usage. Sometimes it is the rules for certain key words, 'not', 'are', 'is', 'all', 'some', 'more': at other times it is the rules for some specialised word; for example, instead of the analytic proposition All red things are coloured, we may have the deductive argument

> This is red
> *therefore* This is coloured

where the validity of the argument, like the truth of the analytic proposition, turns on the meaning of the words 'red' and 'coloured'. Similarly the arguments

> He is an uncle
> *therefore* He is not an only child

and

> He is a bachelor
> *therefore* He is not married

turn on the meanings of the words 'uncle', 'only', 'child', 'bachelor', 'not' and 'married'. But the meanings of words are not always clear. Although we often know how to use words we seldom know how they are used, in the sense of being able to give explicit definitions of them; indeed, we find it very hard to formulate adequate definitions, and normally we just rely on our unformulated "know-how". We are none the worse for that, so long as usage is clear and we agree about it. But often usage is not clear, and we are not agreed about how to use a crucial word. In the United States at one time to say 'He is red' meant that he had left-wing sympathies and to say 'He is coloured' meant that he numbered Negroes among his ancestors. But the inference from the former proposition to the latter is not deductively valid. Again the word 'uncle' is used in England not only of a parent's brother but of any middle-aged friend of the family, and also of pawnbrokers. It is not clear whether these peripheral uses of the word 'uncle' are standard ones or not. Certainly, we could not convict a man who used the word 'uncle' in either of these senses of not knowing English, although if he refused to distinguish the different senses of the word, communication would break down. But the criterion for distinguishing senses is simply the validity or invalidity of various patterns of deductive argument, and so our appeal is no longer to a common language but simply to a willingness to accept certain types of argument as valid. If the only cogent arguments were deductive ones, it might not matter making a willingness to accept them a precondition of communication: but since there are other arguments, many of which are felt by many people to be cogent, we shall be arguing in a circle if we explain deductive arguments in terms of linguistic usage, and linguistic usage in terms of deductive argument.

Language cannot be separated from the rest of life and thought. The meaning of the word 'gentleman' depends on an understanding of the considerations a gentleman is guided by. We speak the same language only because we share to some extent a common life and a common standard of rationality. We often say of a person who completely rejects our assumptions or refuses to acknowledge the cogency of any of the arguments commonly regarded as cogent that "he does not speak the same language as we do". Our language is not a formal logistic calculus but something much more flexible, often vague, sometimes shifting, impregnated with implicit assumptions, something almost alive. Sometimes its shifts reflect merely a social change. At one time the argument

<u>This person is a bachelor</u>
therefore This person is a male

was a valid deductive argument, because unmarried females were called spinsters. But now we have the term 'bachelor girl' and the validity of the argument is in doubt. The meaning of the word 'bachelor' is shifting, and now has connotations of bed-sits, doing for oneself and cooking for one, rather than of being as yet uncommitted and fancy-free. There may even come a day when the word 'bachelor' is so completely anchored in, say, a residential context, that from someone's being a bachelor it will no longer follow as an inference of deductive logic that he or she is unmarried. More serious, from our point of view, than social changes are intellectual ones. Our thoughts change, and our theories have to accommodate new insights or new facts. At one time

<u>This is gold</u>
therefore This is soluble in *aqua regia* but not in *aqua fortis*

would have been a reasonable argument, but not a deductive one, at another it might have been deductive, and at yet another it might have no longer been deductively valid. As our knowledge of chemistry has deepened, and the concepts of alleomorph and isotope have been introduced, the meaning of other concepts has been changed too. The change is sometimes gradual, and we cannot say whether some inference is or is not valid in consequence of the way we use words or in consequence of the way, we believe, the laws of nature operate. Similarly in moral arguments the distinction between moral and purely deductive arguments is sometimes difficult to draw. A man who says that it is always wrong to kill people but that capital punishment is quite all right is clearly failing to use the word 'always' correctly. But is a person who says 'I promise to marry you' and later denies that he ought to marry her failing on a point of linguistic usage or only of morality? If he gave some reason for not carrying out his promise, we should acquit him on the linguistic charge, even if we regarded his reasons as inadequate for going back on his obligations. But if he failed to see that saying the words 'I promise' put him under any obligation whatever, we might also say that he did not know what the word 'promise' meant. There are many other words -- justice, mercy, duty, love -- whose meaning is not constituted by any particular patterns of inference but rather by the recognition of the force of some lines of argument. We cannot separate considerations of language from those of reasonableness. A man who never sees reason to do justice or love mercy, altogether lacks these concepts and does not really understand the meaning of the words.

These difficulties tell not against our drawing the distinction between deductive and other arguments but against the claim that the distinction is always a clear one. Often we can draw the distinction, and it is useful to pick out those arguments which are basically verbal since they turn on the meaning of words, from other more substantial arguments. But we cannot always draw the distinction, and therefore in practice cannot wield simple informal deductive arguments as effectively as we should like, to force recalcitrant reasoners to concede conclusions.

§2 'Not' and 'And'

One natural response is to formalise. We pick out certain basic patterns of inference, and spell out explicitly what the essential pattern is. We choose basic patterns which are, beyond doubt, purely deductive. And if anyone denies their validity, he can be convicted by appeal to the explicit rules of formal logic. We might lay down that the following forms of argument are valid

> All Bs are C
> <u>X is a B</u>
> *therefore* X is a C

and if ever our opponent refuses to accept an argument of this form, we no longer need appeal to his unformulated sense of the meaning of the words 'all', etc., but can simply point out that the argument in question is of this form, and therefore must be valid.

Forms of valid inference have not only been specified explicitly, but have been systematized, so that granted only a few, simple, patterns of inference, others can be validated by a succession of simple steps. Deductive logic is thus very like geometry. At first, in Aristotle's day, there were a number of separate patterns of inference, each recognised as valid, but only a few having been "reduced" to simpler forms. Now, however, there are many axiomatizations of formal logic with only three axioms, some with only one, and correspondingly few rules of inference. The difficulty is that a recalcitrant listener cannot be forced to accept cogent deductive inferences in these formal logics on pain of self-contradiction: If I fail to concede that if you have established the two premises p and $p \rightarrow q$, I cannot deny you the conclusion q, I do not thereby show myself ignorant of English. Rather, I am failing to play the game. The analogy is with cricket: it is as though I refused to leave the wicket after having been bowled. A person who refuses to leave the wicket when out is just not playing cricket; there is no law against it -- the pitch may be in his own garden -- but nobody will play with him. The sanction is the same: being ignored by others if we ignore the rules of the game. Formal deductive arguments thus appear as a species of rule-observance: we don't have to observe rules, but if we do not, we are not observing them. But cricket is optional, whereas logic is not. We could change the rules of cricket, and have bowlers bowl eight times in an over, instead of only six, but we should be chary of recognising "Australian logic" as an equally good way of arguing, simply because the Australians chose to adopt it. Although there is some room for dispute at the margins, the main body of logic is what it is for good reason. Its rationale arises from the need to avoid self-contradiction.

If I have *not both (p and not q)*, or *not p without q*, then if I am given p, I can infer q, because else I should have *p and not q* as well as *not both (p and not q)*, which would be an evident contradiction. We can express this in symbols. [3] Granted that if we have p and $\neg q$, we have $(p \ \& \ \neg q)$, it follows that if we have $(p \ \& \ \neg q)$ and $\neg(p \ \& \ \neg q)$, we have $(p \ \& \ \neg q) \ \& \ \neg(p \ \& \ \neg q)$ which is evidently a self contradiction. That is, if I have $\neg(p \ \& \ \neg q)$, then I can infer q from p, or granted $\neg(p \ \& \ \neg q)$, I have 'if p then q'. This suggests that we write $\neg(p \ \& \ \neg q)$ as $p \rightarrow q$: $p \rightarrow q$ is called material implication, and is often read as 'p implies q'. But it is important to note that in most ordinary uses of 'p implies q' there is a suggestion of some connexion between p and q justifying the inference.

Material implication does not carry this connotation. We can see this if we note that $\neg(p \ \& \ \neg q)$ is true automatically if p is false, so that if we know that p is false, we can assert $p \rightarrow q$ whatever q is, and without there being any connexion between p and q. Similarly if q is true, $\neg(p \ \& \ \neg q)$ must be true too, so that $p \rightarrow q$ holds vacuously. Material implication does not give the full sense of *if, then* but only the minimal truth conditions it must satisfy. But these are important. They are *the* minimal conditions, and if an implication does not satisfy them, it cannot be an implication as we know it. We can express the force of material implication by the inference it legitimises. If I have p and $p \rightarrow q$, then I can conclude q, or, representing inference by $\vdash$, p, $p \rightarrow q \vdash q$.

In reaching this formulation, we have implicitly appealed to inference patterns expressing the force of 'and', $\&$, namely that p, $q \vdash p \ \& \ q$. The complete specification of the logical force of $\&$ has, in addition to introduction rules, the elimination rules: $p \ \& \ q \vdash p$ and $p \ \& \ q \vdash q$. It follows that $p \ \& \ q \vdash q \ \& \ p$. Here, as with material implication, the logical symbol carries only part of the sense of ordinary language. $p \ \& \ q$ means the same as $q \ \& \ p$, whereas "They got married and had a baby" means something different from "They had a baby and got married".

Negation, $\neg$, is similarly defined by inference patterns. In addition to the ban on self-contradiction, it is commonly held that $p \vdash \neg\neg p$ and $\neg\neg p \vdash p$, though the latter inference is controversial. Some logicians work with an "intuitionistic logic" in which only $\neg\neg\neg p \vdash \neg p$ holds. Such a logical system can be developed without inconsistency, but fails to capture the two-sided nature of contradiction. If I say p, you can contradict me by saying not-p, $\neg p$, but I can then contradict you by saying $\neg\neg p$, thereby re-asserting my original claim. Although logical systems can be devised with only a truncated form of negation, or, indeed, without any form of negation whatsoever, they fail to capture the full force of dialogue, in which assertions on the one hand can be confronted by denials on the other, followed by a ding-dong of argument in which claim and counter-claim are made and defended.

Just as we can define $\rightarrow$ in terms of $\neg$ and $\&$, so can we similarly define $\&$ in terms of $\rightarrow$ and $\neg$. We can also define, in terms of $\neg$ together with $\&$, or $\rightarrow$, v, the inclusive 'or', often now written 'and/or', in Latin *vel*, as opposed to *aut*. Similarly we can define $\leftrightarrow$ 'if and only if', and the exclusive 'or', that is 'either or , but not both' in Latin *aut*, in terms of $\neg$ together with $\&$, or v, or $\rightarrow$. [4]

§3 Gödelian Arguments

Formal logic saves us from being swamped by the fluidity of ordinary usage. We may disagree about whether 'or' should be construed in an inclusive or exclusive sense, but the rules for inclusive (*vel* in Latin) and for exclusive (*aut* in Latin) are unambiguous. It would seem natural, therefore, in carrying out the programme of formalisation, to seek total explicitness, with all the agreements expressed in propositions, so that everyone could know that they had been agreed, and nothing depending on an implicit know-how.

But that cannot be achieved. As Lewis Carroll pointed out, [5] we cannot replace all rules of inference by propositions, but need at least one rule, telling us how to operate with the propositions we are given. It is not enough to have the explicit statement

$$((p \rightarrow q) \ \& \ p) \rightarrow q$$

which expresses the truth enshrined in the rule of inference

$$p \rightarrow q$$
$$\underline{p}$$
$$q$$

We need also to be able to *apply* it, which involves being able to recognise symbols, and to see *when* we have premises of the form $p \rightarrow q$ and p, and then *allow* in *that* case that q also holds. There was implicit appeal to a rule of inference in the previous section, in recognising $(p \ \& \ \neg q) \ \& \ \neg(p \ \& \ \neg q)$ as a contradiction of the *same form* as $p \ \& \ \neg p$. Formally, the recognition is justified by a rule of substitution, which is taken for granted in all formal systems. Formal logic is concerned with formal patterns of inference, and has to be able to classify different particular inferences as being of the same general type. Know-how cannot be entirely reduced to know-that: there remains an ineliminable element of know-how: logic is ultimately about how to reason, not a set of formulae in a calculus.

None the less, by formalising we hope to reduce the element of know-how to a minimal ability to recognise shapes, and we can go a very long way in formalising logic in terms of a finite number of axioms and rules of inference. Rules can be applied over and over again. There is a potential infinity of applications. We begin to go from simple deductive logic to mathematics.

Mathematics has long been regarded the paradigm of rigorous deductive reasoning. Many philosophers have sought to cast all reasoning into an incontrovertible chain of rule-validated inferences *more geometrico*, supposing that only so could they establish thought on sure foundations. But much to everyone's surprise, mathematics itself has shown that such an ideal is impossible. This highly unexpected result was proved by Gödel in 1931. [6] Granted that our formal system contains "elementary number theory" in which we can represent both addition and multiplication – it would be a very defective one, if we could not – we can prove in the system that, provided it is a consistent system, it will contain some formula which can be neither proved nor disproved in the system. When we examine the proof we come to the even more surprising conclusion that the proposition expressed by that formula must in fact be true. The Gödelian formula is considered both as an abstract formula in an abstract system, and as an arithmetical proposition with its ordinary meaning. Looking at it in the former way, we prove that, granted the consistency of elementary number theory, it is unprovable: and looking at it the latter way we see that it is true. Thus we have found a new way of establishing the truth of certain propositions of elementary number theory which could not be proved within the standard axiomatization of the theory.

Gödel's theorem is a variant on the Epimenides – or "liar" – paradox, "this statement is untrue", in which the wide-ranging 'untrue' is replaced by a tightly defined 'unprovable-in-the-given-system'. Provided the system is a formal system, and contains enough simple arithmetic to define addition and multiplication, we can code the concept of being 'unprovable-in-the-given-system' into an arithmetical formula, and assign numbers ("Gödel numbers") to each formula of the system. It is then possible to find a number (in practice enormously large, but here pretended to be 1024) which is the Gödel number of the formula 'the formula with the number 1024 is "unprovable-in-the-given-system" ', where the property of being 'unprovable-in-the-given-system' has been given its appropriate code. It follows that the formula no.1024, which we shall call G, must be true, but unprovable-in-the-given-system, since if it were false, it would be provable-in-the-given-system, which would mean that elementary number theory was false.

The actual argument is enormously long, and needs to be stated with great care. But the upshot is that if we formalise any system rich enough to contain ordinary arithmetic, there will be some formula (in fact an infinite number of them) which cannot be proved in it but is obviously true. An obvious retort is that if the proof is valid, it can be formalised, so that a heuristic, informal proof at one stage can be replaced by a proper, formally valid one in due course. It can. But Gödel's theorem applies to *any* formal system strong enough for elementary number theory: and if we try to complete such a system by adding the Gödelian formula as an additional axiom, we can apply Gödel's theorem to this new system to find a new Gödelian formula which is unprovable in the new system. Even if we add a new rule of inference, which would be equivalent to adding an infinite sequence of axioms, we still should not have secured a system that was complete, because Gödel's theorem would again apply to the new system, with its new rule of inference, and would again enable us to find a formula which could not be proved or disproved in the new system, although we would be quite clear that in fact it was true. Gödel's theorem is difficult to understand, easy to misunderstand. The formal proof shows only that *if* the system is consistent, then the Gödelian formula is unprovable in the system. Some philosophers have suggested that the system itself is inconsistent. But that is a counsel of despair; and if elementary number theory were inconsistent, deduction would cease to be a paradigm of valid argument. Other philosophers stick their feet in and refuse to recognise the Gödelian proposition as true. And of course if a man refuses to concede as true anything except what can be proved in a formal system, he can avoid any further embarrassment, beyond that of having to be remarkably obdurate to cogent mathematical reasoning. Essentially what he is doing is to deny any application of the word 'true'. He can understand what the word 'provable' means, and can be forced to acknowledge that in a consistent formalisation of elementary number theory there must be formulae which can be neither proved nor disproved in the formalised system. But he refuses to understand what the word 'true' means, and see that the proposition expressed by the Gödelian formula must in fact be true.

But we do understand what 'true' means, and we learn from Gödel's theorem that truth outruns provability. That has always been believed by some thinkers, though denied by others, who have shied away from any concept of truth that could not be established against the sceptic by some copper-bottomed proof. But once the concept of proof has been made explicit, and the criteria for being provable clearly laid down, Gödel's theorem shows that there are truths which go beyond that concept of provable.

We are led, therefore, to reject certain minimalising views of truth and reason. It is possible for propositions to be true, even though we cannot verify them. It make sense to claim truth, to wonder about truth, to seek truth, beyond the limits of assured knowledge. Equally with reason, to be reasonable is not just to be in accordance with a rule. Aristotle sometimes talks of *kata ton orthon logon*, in accordance with correct reason, sometimes of *meta logou*, with reason; Gödel's theorem underlines the difference, and extends the point made by the tortoise to Achilles in the previous section: however carefully or fully we specify rules of inference, not only is inference something different, but it is not always just rule-observance and can go beyond mere conformity to rules.

Many philosophers have sought to formalise inference. Rather than have inchoate appeals to reason, they seek a few rules of inference which can be precisely formulated, or better still, articulate them as axioms, which can act as premises for simple applications of the *Modus Ponens* rule. Formalisation along these lines can be useful. If the inference itself is in dispute, it helps to try and formulate it precisely, so that each party can spot hidden assumptions, or unjustified claims made by the other. And treating a system purely formally enables important questions of consistency, completeness, or independence to be asked and sometimes answered. But the programme – we may call it "deductivism" – of always formalising every inference cannot be carried through. Deductivists in moral philosophy want to replace every moral inference by a moral major premise, which will apply syllogistically to a proposition describing a particular case. In the philosophy of history, they construe particular inferences about particular situations as covertly universal propositions saying that whenever a situation of the one sort arises, it will be followed by a situation of the other sort. These rational reconstructions always were implausible. Now that we see that the programme cannot in any case be carried through, we need feel no compunction in not formalising in cases where there is no special reason for doing so.

If reason transcends rules, we need to alter many of our views of rational activity. Rules will still be important, but not all important. They provide a useful check on the fallibility of individual reasoners, and a means of agreeing among ourselves about what may be commonly taken for granted. But they need validation by reasonable men recognising that the rules are in fact reasonable, and they are open to criticism. If to be reasonable was simply to follow a rule, then it would be self-contradictory to hold that a rule was unreasonable; but it is always intelligible to say that, and sometimes correct.

Rules should be evaluated at the bar of reason: reason need not necessarily be called to account in terms of rule-following. Often, of course, we are concerned with a rule-governed activity, but not of necessity in all cases. Judgement may be called for, and we may have to decide whose judgement is most to be respected in cases where the rules conflict, or have run out, or do not exist. Although there are disciplines, such as mathematics or mathematical logic, where the traditional aim has been to prove everything *more geometrico*, we do not need, nor should attempt, to reconstruct other disciplines on that model. In history or literary criticism it may be right to recognise the authority of great historians or sensitive critics without being able to reduce their reasoning to explicit syllogistic form. We often talk of originality and creativity in artistic work, and hold that the great artist breaks out from the canons of correct taste, and achieves something that, although not conforming to them, is nevertheless absolutely right: Gödel's theorem underwrites this possibility, as one obtaining not only in artistic creation, but throughout the whole realm of rationality, even in the austere field of mathematical logic.

§4   Analyticity

Gödel proved another theorem, a completeness theorem. He proved that first-order logic is complete, that is that every well-formed formula in first-order logic is a theorem. This is most easily understood if we consider propositional calculus. Propositional calculus can be treated axiomatically, as it was by Russell and Whitehead, who postulated five axioms and certain rules of inference, and proved well-formed formulae by deriving them from the axioms by means of the rules of inference. But we can also use truth tables to work out the truth-value of a well-formed formula from the truth-value of its components; and if a well-formed formula comes out true whatever the truth-values of its components, we call it a tautology. It is a fairly easy exercise to check that the axioms are tautologies, and that the rules of inference are such that if the premise or premises are tautologies, then the conclusion must be one too. It is much harder to prove the converse of this, but it can be done: every tautology can be proved from the axioms. We say that propositional calculus is *complete* (or sometimes "semantically complete"). Gödel proved that the same holds good for first-order logic, the logic in which we have quantifiers, All, Some, None and Not All, with the quantifiers ranging over individuals, so that we can talk of All men being mortal, and the like. We can construe this as showing that the quantifiers, *when they range over individuals*, are subject to the same discipline as we have worked out for And and Not. If you deny the conclusion of some syllogism that holds whatever individual cases it is about, you are not only guilty of contradicting yourself, but can be brought step by step to a self-evident self-contradiction. This supports the claim by Kant and many modern philosophers that deductive logic is purely analytic, based on the principle of non-contradiction, and yielding only empty tautologies.

But Gödel's completeness theorem holds only for *first*-order logic, in which quanification is allowed only over individual variables, and not over predicate variables. In first-order logic we can talk about all Peter's children, and say that they have blue eyes, but not about all Peter's features, which none of his children possess in their entirety. Dedekind needed the latter sort of locution in order to define the natural numbers, and it is similarly needed to define the concept of being finite, and to formulate various different axioms of infinity. Second-order logic allows quanification over predicate variables, so that we can talk about all features, all properties, all qualities, and the like, but second-order logic is *not* complete in Gödel's sense, but only Henkin-complete, which secures completeness only by misinterpreting key logical terms. In second-order logic not all well-formed formulae that are true under all (reasonable) interpretations are theorems. Logical truths are not always provable by a step-by-step ("algorithmic") derivation. A sceptic can deny them without being led into patent self-contradiction. Deductive logic is not necessarily analytic: synthetic *a priori* propositions are possible.

And we need to have them.

§5   Mathematical Dialogues

The argument of the previous section is cogent, but not incontrovertible. All that has been formally proved in any system which contains simple arithmetic is that there is no formal proof in the system of the Gödelian formula or its negation. A determined sceptic could maintain the negation of the Gödelian formula without contradicting himself – just. We could argue with him – there are cogent arguments for not accepting the negation of the Gödelian formula – but his position, though weird, is tenable: we can understand what is being claimed; it is not a contradiction in terms, and if put forward, would need to be argued against, not simply dismissed as unintelligible. It follows that even this austerely mathematical reasoning is two-sided. And, contrary to orthodox opinion, much else in mathematics is best understood in terms of a two-sided dialogue between seekers after truth.

Mathematics, with its unexpected incompleteness and unprovable consistency, is different from simple, first-order logic, which is provably consistent and complete. Poincaré [7] regarded the Rule of Recursion, often misleadingly called the Principle of Mathematical Induction, [8] as the key difference. In Peano's first-order axiomatization of arithmetic, it is expressed by the axiom schema

$$(F(0) \ \& \ (\forall n)(F(n) \rightarrow F(n+1))) \rightarrow (\forall x)F(x)$$

which lays down that if the property F holds for 0, and is such that if it holds for any one natural number it holds for its successor, then it holds for all natural numbers. It seems a very reasonable rule to adopt. It does not seem to add anything that could be questioned to what we already have. If we had Peano's other four postulates without this axiom schema, we should have a system that could be called "*Sorites* Arithmetic", in which, granted the antecedents of the axiom schema, namely

$$F(0) \text{ and } (\forall n)(F(n) \rightarrow F(n+1)),$$

we could establish for each natural number n that $F(n)$. There could not be a counter-example within what we normally regard as the natural numbers. If anyone were to maintain that there was, and that for some number, say 257, it was not the case that $F(257)$, we could construct a *Sorites* argument

$F(0)$

$F(0) \rightarrow F(1)$ (particular case of $(\forall n)(F(n) \rightarrow F(n+1))$

$F(1)$ by *Modus Ponens*

$F(1) \rightarrow F(2)$ (particular case of $(\forall n)(F(n) \rightarrow F(n+1))$)

$F(2)$ by *Modus Ponens*

$F(2) \rightarrow F(3)$ (particular case of $(\forall n)(F(n) \rightarrow F(n+1))$)

$F(3)$ by *Modus Ponens*

$F(3) \rightarrow F(4)$ (particular case of $(\forall n)(F(n) \rightarrow F(n+1))$)

$F(4)$ by *Modus Ponens*

. . .

. . .

$F(256)$ by *Modus Ponens*

$F(256) \rightarrow F(257)$ (particular case of $(\forall n)(F(n) \rightarrow F(n+1))$)

$F(257)$ by *Modus Ponens*

Thus in the end he would be forced to withdraw his putative counter-example, on pain of self-contradiction. In fact, he could see that this would be the outcome long before we had finished the proof, and would concede rather than waiting to be checkmated. And he would see that the same would happen if he were to suggest any other putative counter-example. It would be reasonable, therefore, for him to concede the claim that $F(n)$ holds for all n.

But we have not actually proved that. We have shown how, for each n, we can prove $F(n)$; but without the Rule of Recursion or some equivalent, we cannot actually produce a formal proof, the last line of which is

$$(\forall x)(F(x))$$

expressing the proposition that *all* natural numbers x have the property F. There is a subtle difference between being able to prove, given any particular natural number n, that F(n) holds, and actually proving that all numbers possess the property in question. Nevertheless, we are confident of the validity of the Rule of Recursion, and generally regard arguments invoking it, along with the rest of mathematics, as deductive.

But there is a difference. The difference shows up in a difference of sanction. The sanction against those who refuse to recognise the cogency of arguments by recursion is no longer that they are contradicting themselves or failing to abide by the rules of calculus they are operating. The Rule of Recursion is independent of the other axioms and rules of inference of first-order Peano arithmetic (because our concept of natural number is not sufficiently tightly formulated to exclude the possibility of there being some "inaccessible" numbers, which could not ever be reached by a chain of arguments, no matter how long they were). Someone who refuses to concede that every number has the property F, granted that 0 does, and that if any number does, so too does its successor, cannot be convicted of a straight contradiction, however unreasonable his stance seems, provided he sticks to his position without making any further move. Although it is very natural to give reasons for denying a claim made by a friend, it is not absolutely obligatory, and an antagonist who plays with his cards very close to his chest can avoid being brought up short in a self-contradiction. But once he starts considering the possibility of counter-examples, he is lost.

Why should he be reasonable? Or better, Why should I be reasonable? If my sole aim is the polemical one of not losing the contest, there is no reason why I should expose myself to being worsted in the argument. But the case is altered if I want to know the truth. In that case I value discovering what actually is true more than winning the argument, and indeed would rather be shown wrong in what I had previously believed, if by that means I come to exchange wrong opinions for better ones. If I have any idea of truth, it is something that my beliefs should conform to, rather than something that should be conformed to what I think. Truth is independent of me, or it is nothing. And however good an opinion I have of my opinions, once I am possessed of the idea of truth, I know that the opinions which are worthy to be believed are the true ones, not those that happen to be believed by me.

A reasonable man, therefore, will make the moves necessary for ascertaining the truth, even if he thereby exposes himself to being refuted. He will, "for the sake of argument", suppose that some particular natural number was a counter-example to the thesis being put forward, and, then seeing that any such supposition would involve him in inconsistency, allow that no counter-example could exist, and hence that the thesis must be true. The love of truth makes one vulnerable to pressures to which the purely contentious man is immune. Although the sanction against rejecting the Rule of Recursion is different from the sanction for rejecting *Modus Ponens*, arguments invoking recursion are properly regarded as deductive.

§6 All, Any, Every and Each

The terms 'all', 'any', 'every' and 'each', normally interchangeable when we are talking about the finite cases of ordinary converse, reveal logical differences when we are deploying formal proofs over infinite domains.

'All' is the strongest claim. Once admitted, it applies universally without more ado; I defy you to produce a counter-example; and – if my claim is admitted – you know you cannot. But before it has to be admitted, it needs to be justified to the hilt. In order to justify the claim 'all . . . ', I must produce a straightforward derivation the last line of which is

$$(\forall x)(F(x)).$$

'Any . . . ' is not so forthright. Instead of saying that I have got a proof, I invite you to choose some particular instance, x, and then I show $F(x)$ by an incontrovertible derivation of a standard form. Since it is a standard-form derivation, I can myself apply it mechanically, and hence instruct you to apply it mechanically, and thus lead from an assured ability to prove any particular case you happen to choose, to a straight proof that my claim does, indeed, hold for *all* cases. *Any* entails *all* when the proof of *any* is always of the same form.

With recursive reasoning, the inference from *every* to *all*, the proof in each case is indeed valid, but not *quite* the same. The number of *Sorites* steps needed depends on the particular example chosen. The proof of $F(257)$ outlined in the previous section was similar to that of $F(4)$, but not *precisely* the same, inasmuch as it had 515 steps, whereas only 9 were needed to prove it for $F(4)$. Similarly the contrapositive disproof of the denial of $F(257)$ is much longer than that for $F(4)$. *Any* entails *all* because

$$(\forall x)(\exists Disproof_{standard\text{-}issue} \neg F(x)),$$

whereas *every* entails *all* because

$$(\forall x)(\exists Disproof_x \neg F(x)).$$

Since the disproof is an incontrovertible derivation, it is reasonable to expect any reasonable person to recognise that fact, and to concede without more ado, but if it is to be carried through to the bitter end, the respondent has to make the necessary adjustments to the length of the derivation, before he sees the inevitable outcome. The cooperation required of the respondent is minimal, but enough to mark the distinction between the inference from *any* to *all* and the inference from *every* to *all*.

Recursion is implicit in our ideal of a formal proof. A formal proof consists in a finite number of steps, each of which is a valid inference, in virtue of some stated rule of inference, of the proposition (or formula or sentence) in question from some proposition(s) which either is an axioms or is already established at some earlier stage in the proof. It must consist of only a finite number of steps, but can consist of any finite number of steps. That is, it can be as long as we please, but not infinitely long. It follows that as soon as a number, any number (but some definite one), is picked, we can prove our formula for that number; but that as long as the number has not been fixed, there is no guarantee that any particular proof we have produced will have reached that number.

The derivations adduced to make good a claim that 'every...', although not precisely the same in each case, are very similar, differing only in the length of the derivation, according to some specifiable formula. The respondent is required to exercise only minimal cooperative intelligence to see how the *Sorites* derivation should be adjusted in length to fit the case under discussion. But it could be that in each case there was a derivation, but not so similar that the respondent could be given simple-to-follow instructions how to tailor the basic model so as to fit the particular case. The Gödelian argument sketched in this chapter can be deployed with regard to any first-order system strong enough to include elementary number theory, but there is no set formula giving instructions how to do it in each case. Some non-minimal cooperative intelligence is required on the part of the respondent to see how the general Gödelian argument should be tailored to fit the particular case under discussion. Just as the argument from *every* requires more cooperative intelligence than the argument from *any*, so the argument from *each* requires more cooperative intelligence than the argument from *every*.

The problem becomes tractable if we use the dialectical approach to cast these arguments in negative form. In the case of 'all . . . ', the negativity arises only from the nature of deductive proof. A formal derivation is a sequence of steps, each one of which must be allowed on pain of self-contradiction. In claiming 'all . . . ', I need to be able to produce a derivation which must be conceded on pain of self-contradiction. That is, if I say 'all . . . ', and you were to say 'not all . . . ', I could lead you step by step to inconsistency. 'All . . . ' is demonstrably ungainsayable.

Positive Universal Arguments

1. All. I have a proof of $(\forall x)F(x)$ and here it is.

2. Any. You choose x, and I will prove $F(x)$ by a standard procedure which will work for any other x you might choose.

3. Every. You choose n, and I will prove $F(n)$ by a *Sorites* derivation which will work, with suitable adjustments, for any other n you might choose.

4. Each. You choose x, and I will prove $F(x)$. You should be able to get the hang of my proof, and see how it could be altered to fit any other x you might have chosen.

–o0o–

Once dialogues (2), (3), or (4), have been concluded, you, if you are rational, will reckon that you cannot fault an All claim. So next time we meet, I claim All.

Whereas in these latter dialogues, I give you the first move, and then persuade you to concede, I now make the first move myself, defying anyone to controvert me. All is a challenge I make to all comers, whereas with Any, Every and Each, I let you have the first move, so that you can see that try as you can, you cannot evade defeat.

With *any*, *every* and *each* the negative approach is much more significant. Instead of my saying that I have got a proof, whereupon you may challenge me to produce it, I challenge you to produce a counter-example, whereupon I show that it is not a counterexample in such a way that you realise that whatever purported counter-example you might propose, I should be able to refute your claim that it really was a counter-example. Provided you take up the challenge, and name some particular purported counterexample, I shall be able to prove you wrong.

If I claim 'any . . . ', I challenge you to find a case that is not ungainsayable. If you deny $F(x)$ thereby claiming $\neg F(x)$, I shall show that $\neg F(x)$ leads to a self-contradiction, so that $F(x)$ is, as I claimed, ungainsayable. And, as argued before, since my proof is in a standard form, applicable without more ado to any case, you can see (and if you cannot see, I can tell you) how to assure yourself of the incontrovertibility of $F(x)$, whatever x you choose.

The difficulty with the positive approach to the inference from *every* to *all* was that the respondent had to make the adjustments necessary for the proof to fit his particular case. There was no upper bound to the length of proof that might be required, and since proofs have to be finite, we feel queasy at the prospect of their being indefinitely, even though not infinitely, long. Qualms are avoided in the negative approach, since although we have an infinite progression of numbers, for every individual number our argument can be put in the form of a regress which is necessarily only finite: I can prove my refutation of a purported counter-example for n, provided I can prove it for n-1; and I can prove it for n-1 provided I can prove it for n-2; and so on, until I come down to 0, where I have already established it. The dialectical challenge reverses the direction of the burden of proof. The burden of proof is still on my shoulders, but instead of my taking on the impossible task of scaling an infinite ascent and proving my claim for all of infinitely many cases, I get you to propose to me the manageable task of descending from any given stage and showing how well-grounded my argument is there. And this I always can do, in a finite number of steps. For although the natural numbers go on without ever coming to an end, they do have a definite beginning. And the Rule of Recursion secures its credit by trading on this fact.

The method of argument can be extended to transfinite numbers, since every transfinite ordinal can generate only a finite chain of descending ordinals. Transfinite set theory often appeals to the Axiom of Choice, which can again be illuminatingly viewed in terms of a dialogue, or alternatively, to the Axiom of Determinacy, which is explicitly dialectical in form.

In the straightforward positive justification of the Gödelian argument some significantly non-minimal cooperative intelligence was required on the part of the respondent, and if this is not forthcoming, it cannot be forced on an unwilling respondent. Hence again the polemical value of casting the argument in the negative form, where the respondent chooses some particular system, and I point out its Achilles' heel.

Negative Universal Arguments

1. **All.** If you deny $(\forall x)F(x)$ I shall show that $\neg(\forall x)F(x)$

leads to an inconsistency, so that

$(\forall x)F(x)$ is ungainsayable.

2. **Any.** You choose x, and if you deny $F(x)$, I shall show that $\neg F(x)$ leads to an inconsistency. I shall show it in a standard way, which clearly would apply to any x you might have chosen.

3. **Every.** You choose n, and if you deny $F(x)$, I shall show that $\neg F(n)$ leads to an inconsistency, by a step-by-step contrapositive *Sorites* argument, ending with a denial of $F(0)$, which was one of the premises originally given. You should be able to twig what is going on, and realise that whatever $F(n)$ you propose, I shall refute you, but if you do not, I can spell it out for you. If you cannot hoist it in, you can go on trying to produce a counter-example, and failing every time.

4. **Each.** You choose x, and if you deny $F(x)$, I shall show that $\neg F(x)$ leads to an inconsistency. You should be able to get the hang of my proof, and see how it could be altered to fit any other x you might have chosen; but if you cannot, you can go on trying to produce a counter-example, and failing each time.

§7   Induction

Induction perplexes philosophers. Inductive arguments are not valid deductive arguments, yet clearly possess some cogency. It is difficult to maintain a position of complete scepticism about them, but impossible to justify them by purely deductive means.

Many different characterizations of induction are given: the inference from the known to the unknown; from the past to the future; from the particular to the general; or from a certain number of particular instances that we have observed to another particular instance that we have not observed. These characterizations are not equivalent, nor does any constitute a definition. Not only do they pick on different aspects of inductive argument, but inductive arguments are of different types: sometimes posing different difficulties to the sceptic and requiring different sorts of justification. Nor is any one type adequately defined. Indeed, it is difficult to define induction. It is easy to begin, and lay it down as a necessary condition that inductive arguments are not deductive. An older generation of philosophers used to divide all argument into deductive and inductive, and define inductive arguments as those that were not deductive. Hence the characterization of the Rule of Recursion, which was recognised as not being a simple deductive argument, as Mathematical Induction. Some philosophers would still stop there, though with a more adequate characterization of deduction, and offer, as a complete definition, that an inductive argument was a non-deductive one. But evaluative arguments, moral arguments, political arguments, legal arguments, historical arguments, literary arguments, philosophical arguments and theological arguments, are nearly all non-deductive, but very far from being all inductive. Some of these would be excluded if we defined inductive arguments as empirical arguments, or as arguments about matters of fact, but our idea of the empirical is hazy, and the concept of a fact is systematically ambiguous, varying very much with context. [9] Even if philosophy, theology, and literary criticism are deemed non-empirical and non-factual, history seems to be concerned with facts, and depends on empirical evidence, though few of its arguments conform to the canonical pattern of induction. As a rough criterion, we could say that inductive arguments are concerned with things rather than persons, and that anything evaluative or interpretative should be excluded. Some philosophers have attempted to exclude unwanted cases by stipulating that the conclusion of an inductive argument should be of the same logical type as the premises: if the premises are propositions about individual swans being white, so should the conclusion; from 'is's other 'is's may be derived inductively, but not 'ought's. Although some types of inductive argument can be adequately defined in this way, others cannot: for example, the argument from particular instances to general laws, which was historically the first type of inductive argument to be distinguished. None of these definitions is satisfactory: like deduction, induction is fuzzy-edged. It is best to start with widely accepted standard cases, and recognise that the decision as to which other cases should be included is somewhat arbitrary, to be settled on grounds of convenience as much as anything else.

Induction

1. Inference to the next case
2. Inference to the general case
3. Inference to natural law
4. ?Inference to best explanation?
   (also called abduction, to distinguish it from the first three)

Let us start, then, with the simplest case, where we argue from particular to particular, and then extend it, in a natural but highly revealing way, to the argument from particular to general. In a simple induction we argue from a number of particular instances of a specified sort and having some further characteristic to some other particular instance's having the same further characteristic. The argument can be displayed crudely and inadequately by the following schema:

> This swan is white
> That swan is white
> A third swan is white
>
> …
>
> …
>
> A 256th swan is white
> *therefore*   The 257th swan is white.

We may call this type of inductive argument Inference to the Next Case. Other examples are the argument that the sun will rise tomorrow, [10] and Russell's chicken which on the basis of observation concluded that the farmer was coming to feed her on the day he was coming to wring her neck. [11] Inference to the Next Case is the type of inference that Hume had in mind, when he sought to explain it in terms of a conditioned reflex. Admittedly, reflexes can be conditioned; but Hume's contention that it is just a matter of habit is not plausible. For one thing, we sometimes cite reasons why a conclusion is to be believed, not just causes explaining how we have come to hold it; for another, we often jump to conclusions far more quickly than any habituation process could take place; often we bring to bear a whole lot of background information which justifies our taking a particular observation as indicative of the way things are, without needing to repeat it again and again.

Inference to the Next Case

The Sun will rise tomorrow (note danger of tautology)
The Next Swan will be White
Russell's chicken – *The Problems of Philosophy*, p.63
Is it just a matter of habit, as Hume holds? No, because:
  (a) we give reasons, saying not only why we do, but why one should, accept the conclusion
  (b) we jump to conclusions without waiting to be habituated to them

Inference to the Next Case leads naturally to Inference to a Generalisation, or Inductive Generalisation, as it may be called. The schema of argument given above will yield conclusions not only about the 257th swan, but about the 258th, 259th, and indeed about any particular swan. Hence it is natural, much as in the case of the Rule of Recursion, to state the form of argument not as a schema that could be applied to yield a conclusion about any particular instance, but as a single argument yielding a conclusion about all instances, *i.e.* a conclusion in universal terms. This is the traditional form

> This swan is white
> That swan is white
> A third swan is white
>   …
>   …
> <u>A 256th swan is white</u>
> *therefore*   All swans are white.

The conclusion is of a different logical type from the premises. Its universality is derived from the universality of reason. In simple induction we had a pattern of valid reasoning, which therefore applies in any particular case, and so is of universal application. And this is made explicit in the universal form.

As far as particular cases go, Inference to the Next Case and Inference to a Generalisation come to much the same thing. Granted the former. we can establish that something holds for each and every case, and granted the latter, we can obtain the particular case by one further step of deduction

<u>All swans are white</u>
therefore The 257th swan is white.

Nevertheless, the difference in logical type of the conclusion is of considerable philoscphical importance, and brings to light covert assumptions in drawing the inference, as well as possible justifications of the whole pattern of argument.

The pattern of argument thus far displayed is defective. It is not good enough to argue

This swan is white
That swan is white
A third swan is white

...

...

<u>A 256th swan is white</u>
therefore All swans are white.

The premises, as it happens, are all true, but I do not draw the conclusion because I have heard tell of black swans, some growing naturally in Australia, others more accessibly visible in Chartwell. In order to reach the conclusion I need a further premise

I have never seen, or heard of, a swan which was not white

This is expressed more traditionally

<u>All the swans I have ever seen were white</u>
therefore <u>All swans are white</u>
therefore The next swan will be white

in which the special force of the additional premise is played
down. But this is a mistake. It plays down the two-sidedness
of inductive inference, the importance for looking for
arguments against the conclusion, and only accepting the
conclusion after a reasonably sustained search has thrown up
nothing substantial. [12] The importance of the additional
premise is brought out in Popper's approach to inductive
inference. Popper does not ask how the conclusion may be
verified, but how it may be falsified, and since a single
counter-instance, unless it could be explained away, would
be enough to falsify it, Popper's main concern is to look for
counter-instance, and to regard any universal proposition that
has not been refuted as a candidate for truth. [13] As soon as a
black swan turns up, I abandon the claim that all swans are
white: but until the existence of such a swan is brought to my
notice, the law that all swans are white is a reasonable one;
and therefore the critical issue is whether I have ever come
across an un-white swan or not. Although Popper's account
explains the importance of the additional premise, it does so
at the cost of making all the others seem unnecessary. I do
not need to have seen any white swans at all in order *not* to
have *refuted* the universal proposition All swans are white.
Exclusive emphasis on falsification does not accommodate
our everyday belief that positive instances are as important
as negative ones, and that a large number of positive
instances does, in the absence of any negative ones, increase
the evidence in favour of the conclusion. Nevertheless, even
though it errs in neglecting the need for arguments in favour

of any putative conclusion, Popper's account is valuable in stressing the importance of considering arguments against.

Although Inference to a Generalisation is inter-arguable with Inference to the Next Case, its conclusion is of a different logical type. 'The next swan is white' is not only a particular proposition, but a tensed one. 'Every swan is white', or, equivalently, 'All swans are white', is not only general, but tenseless; we can infer from 'Every swan is white' and 'Leda was a swan' the conclusion 'Leda was white'. Such an inference would not be valid if the 'is' of the first premise were a present-tense 'is'. It is, rather, an omnitemporal use of the verb 'to be' which is put into the present for lack of a better tense to put it into. Such a use of the present tense is sometimes indicated, following a suggestion of J.J.C. Smart, by italics. [14] So we write 'Every swan *is* white' or 'All swans *are* white', to indicate that the grammatically present tense is being used in a logically tenseless way, or, better, use the grammatically incorrect 'Every swan *be* white' or 'All swans *be* white'. Such a use is entirely unobjectionable. But it heightens the profile of induction. Induction does not merely argue from particular to particular in the ordinary tensed indicative mood, but from particular in the ordinary tensed indicative mood to general in a different, tenseless mood. The mood is clearly different, not only because it does not conjugate like the ordinary indicative mood, but because it yields counterfactual propositions, such as 'If Zoe were a swan, she would be white', which the ordinary indicative mood does not.

Inference to a General Case

The sun rises every day. All swans are white.

(a) Importance of absence of negative instances: inductive arguments of this type are essentially dialectical, two-sided: only if I have looked for negative evidence and failed to find it, am I entitled to draw a general conclusion.

(b) Inference to Next Case and Inference to a General Case are interderivable, but differ in logical status of the conclusion. Conclusion of former same logical type as premises (tensed indicative): conclusion of latter in tenseless present (*be*).

Often we go further still, and argue from particular premises not only to an omnitemporal generalisation in the tenseless present, but to a law of nature stating what must, under appropriate conditions, occur. It then becomes difficult to disallow, as also a species of inductive inference, arguments from actual instances to natural laws and from observed phenomena to unobserved entities. We argue from the regular whiteness of swans to a rule that they must be white, and from white appearances to a genetic make-up that accounts for them. Such inferences, although rejected by Hume, have commended themselves to scientists ever since. We seek generality, integration, unification and explanation in our account of the world, and it seems reasonable so to seek. Although quarks, psi-functions and wavicles all transcend the bounds of possible experience, we form some sort of concept of them, and succeed in saying things about them which can be significantly affirmed or denied. Nobody makes out that the Special Theory, the General Theory and

Quantum Mechanics are plain sailing. They are difficult, and it is easy to be confused and talk nonsense about them. But it does not follow that rational argument about science is impossible, or that reason must acknowledge that such knowledge is too high for it, and it cannot attain unto it. The arguments Hume put forward for ruling out altogether knowledge of unobserved entities or explanations of the universe as a whole, would, if they were cogent, rule out all sub-atomic physics and cosmology. But, while many thinkers fear – or hope – they are cogent when deployed against metaphysics or natural theology, few seriously suppose they cast any aspersions on the reputability of modern science.

Inference to a Natural Law

In any isolated system mass and energy must be conserved. The halogens can only have one valency bond.

(a) Importance of a wide variety of efforts to falsify the law: only if I have tried hard to find negative instances and failed to find them, am I entitled to infer that I have come up against a law of nature.

(b) Importance of consilience: natural laws should mesh together.
Inference to a Natural Law yields a conclusion very obviously of a different logical type from premises. It shows us that reason can go from one mood to another: although we cannot *deduce* an 'ought' from an 'is', it may be possible to *derive*, or *infer*, it. This has consequences for metaphysics as well as for morals.
It shows that we can go, with reason's aid, beyond the bounds of possible experience.

It is a moot point whether inferences leading to the acceptance of Scientific theories should be called inductive. In recent years they have often been termed 'inferences to the best explanation'. The term 'abduction' is sometimes used. The name indicates a difference of sanction. If having noticed a number of cases of people drinking hemlock and subsequently dying, I refuse to infer that if I drink hemlock I shall die, I shall pay for my scepticism with my life.

Inference to Best Explanation

An example of inference to entity of very different *ontological* type: Leibniz' argument from *La Liaison des Phénomènes* for the existence of material objects.

The disutility of not reaching right conclusions about matters of fact, especially matters of future fact, is so great that even the most obdurate sceptic finds ways of not living down to his professions of ignorance. Omnitemporal generalisations are also useful, and the sceptic who disallows Inference to a Generalisation is depriving himself of much useful knowledge in a convenient and memorable form. Laws of nature and Scientific theories can also be defended on grounds of utility, but the argument is more tenuous: many effective people lead happy and successful lives without having mastered Einstein's General Theory, or Quantum Electrodynamics. Although some knowledge of science is in some circumstances useful, the real sanction against the sceptic who will not admit inferences to Scientific theories is that he will suffer from avoidable ignorance. I want to know the nature of the universe and to understand the causes of things. Natural laws and Scientific theories claim truth and offer understanding. They integrate and explain. Diverse phenomena are unified by being brought under a single principle, and a welter of confusing events are explained by means of a theory. It is because they unify and explain that we believe that the laws and theories are true. But if they are true, we want to know them. The same motive applies also, though to a lesser extent, to accepting Inference to a Generalisation, and even to Inference to the Next Case. We like to know the way things happen, quite apart from any utility. Few bird-watchers eat birds, or obtain any material benefit from knowing which species the birds in the garden belong to, but they want to know all the same. Although our motives may be mixed, our methods are the same: we apply various principles, often unconsciously, to distinguish good inductive inferences from bad ones.

By contrast to the systematic way in which we can test different combinations of possible causal factors, there is no systematic test for explanatoriness. We have a number of general ideas of what possible explanations there can be, and are much readier to accept a putative causal generalisation or law of nature if we can see how it might be explained – if it fits in with what we already know about the way things happen. Medical scientists are very reluctant to accept evidence in favour of homoeopathic remedies, and try hard to ascribe any indubitable cases to chance, because they do not see how homoeopathy could work. Consilience – the extent to which a new generalisation will fit in with what is already known – is a major factor in inductive inference, and one that cannot be reduced to a set of systematic rules. Nevertheless, all in all inductive inference is a rational activity, largely systematic, rule-governed, and we can properly speak of "inductive logic", as it used to be called.

We are not obliged to accept inductive arguments as we are deductive ones Communication will not break down if I, having seen many white swans, refuse to infer that the next one will be white too. It is perfectly intelligible to maintain that the next sample of hydrogen cyanide will not be poisonous – but highly unwise. Without the aid of inductive inferences I shall not be able to anticipate the situations I shall have to face; nor ward off untoward outcomes of conditions about to obtain. I shall know less, if I do not allow myself to infer universal propositions, under suitable safeguards, from particular observations, and I shall understand less, if I do not go beyond the phenomena to the best explanation available.

§8 Practical Reason

Once it is recognised that inductive inferences can lead from a tensed 'was' to a tenseless '*is*' or '*be*', it becomes hard to maintain as a matter of logical principle that we cannot derive an 'ought' from an 'is'. Even deduction can lead us to conclusions not contained in the original premises. And in Inference to the Best Explanation we argue from premises of one sort to conclusions of a very different logical type. The conclusions of practical reasoning are of a different type again. They are action-guiding. And the premises are factual, characterizing the situation to which the action commended is a response. But if we reason at all about what to do, our conclusions must point towards some action as appropriate, and if our reasoning is to be relevant and effective it must be based on the facts of the case.

The particularism of practical reasoning seems to run counter to its rationality. Rationality is universal, and enjoins us to act only on that maxim through which we can at the same time will that it should become a universal law. [15] This would require that we had a stock of universal laws under which we could subsume any action indicated by practical reason. But any such set of universal precepts would be too coarse-grained to accommodate the unbounded subtlety of human affairs. To accommodate that, we need a more flexible canon of universalisability. Instead of requiring that there be some universal precept that covers the proposed action, and that all other similar ones should be treated similarly, we should require only that *if* we propose to treat some apparently similar case differently, there should be some difference between them to justify the difference of treatment. [16] Granted this different requirement of universalisability, we can have practical arguments being both rational and relevant to the particular case under consideration.

It is the particularity of the situations in which we have to act that requires our assessments to be holistic and gives rise to the two-sidedness of practical reasoning. However fully we have specified a situation, there always may be a further feature that entirely alters the complexion of the case; and since circumstances, too, alter cases, there always may be some extraneous circumstance that requires a re-assessment of our response with a further 'but'. The fact that there is nearly always room for another 'but' has often been taken to show that there are no valid arguments in practical reasoning, but what it actually shows is that there are few *conclusive* arguments. Many arguments are cogent in the absence of counter-considerations, and we often state them explicitly with this proviso, "other things being equal", *ceteris paribus*, "in the absence of special circumstances", "as a general rule", *hos epi to polu*.

But other things may not be equal. That a certain action will cause you pain is a good reason for not doing it, yet I could well go on to say "I am afraid this will hurt, but it is what I have to do": I might be a dentist, or a head-master, or an examiner, or a candid friend. But if I do conjoin 'this will cause pain' and 'this is what I ought to do', I need to explain why I ought to do it, in spite of its causing pain; perhaps because it will promote good health, or teach you a lesson, or uphold the integrity of some public system of evaluation, or enable you to make rational decisions in the face of unpalatable facts. It is something that calls for further explanation. The explanation need not be a moral or benign one: it would not be unintelligible, though it would show me in a bad light, if I explained that I was a sadist, and liked causing pain, or a Nietzschean, and wanted to make my mark on the universe.

Often a consideration is not so much countered as over-ridden. My having promised to return a borrowed weapon is a valid reason for doing so, but if the person I had borrowed it from has in the meantime gone mad, the obligation to return is over-ridden by concern for his, and other people's, safety. But that does not abrogate the original obligation. If for good reason I have had to break a promise, it is not as though the promise had never been made: on the contrary, I am subject to further obligations to make good to the person I let down the damage my broken promise caused. We need to distinguish cases where a putative obligation does not hold at all – as if I made the promise having been deceived by the person to whom it was made – and cases where it is simply over-ridden. [17]

Our decision will depend not only on the strength of the arguments on one side, but on the weakness of those on the other. For only when there are alternative courses under consideration can I decide between them. If you counsel flight, I shall re-examine the situation to see whether it would be better after all to flee than to give battle, and with these alternatives in view I can try and tell whether this is a sticking-it-out situation or a cutting-one's-losses one. If, alternatively, you counselled me to advance, making propitiatory gestures, I should need to re-examine the situation in a different way, to attempt to make a different discrimination. So, again, if you had urged me to go on the offensive myself and launch a pre-emptive attack. The alternatives offered determine the factors that are relevant. I do not have to – very often cannot – justify my own judgement absolutely, but only relatively: I am not bound to show that I ought to do this, full stop, but only that I ought to do this rather than that. Different factors are relevant for deciding between different pairs of alternatives. If you counsel flight, it is pertinent to point out that I cannot run very fast: but my inability to run would be no argument against making propitiatory gestures, though the fact that my adversary was a cowardly bully would. Once the alternatives are fixed, we shall each look for features of the situation which will enable us to discriminate between its being a staying-and-sticking-it-out situation and its being a cutting-one's-losses one. We may be able to: language, used by both of us on many occasions before our present argument, enables us to pick out many sorts of features which might be relevant. We can compare this situation, which we each read differently, with others where we have no disagreement. 'But he looks very threatening', you may say of the adversary, 'Yes, but he is also anxious to secure his retreat', I counter; but because we are using some other, and usually more general, classificatory scheme than our primary one, we can work round to an agreed description of the situation we are considering. We may still be unable to reach agreement about what to do: we may argue that he is looking round nervously; or you may suggest that he is looking round not in order to see how he may beat a hasty retreat, but because he is expecting his accomplices to be on their way to join him. In that case we could agree only that he was looking round rather a lot. Even that description is not immune to objection. You may say that these few casual glances are no more than anyone might cast behind him, to note the terrain covered or to admire the view. In some situations you might question whether the man came with any hostile intent at all, in others whether it even was a man that I saw.

There is no uniform agreed level of facts from which our arguments can start. Rather, what counts as a fact depends on the question in issue. Almost anything may be disputed: but invariably there are some facts not in dispute, and it could not be the case that almost everything was disputed, for then there would be no common language in which to carry on the dispute.

Although we often disagree how the balance should be struck in some particular case, we very largely agree about the relevant considerations. When it comes to making decisions in particular situations, it always may be the case that peculiar circumstances may make it impossible to do all the things that normally ought to be done: but that does not affect the general reckoning that in all ordinary circumstances they should indeed be done. In general the reason is a good reason for acting in a particular way, even though in the exceptional case the reasons on the other side are weightier. Apart from a few disputed issues – abortion, capital punishment, sex – we all allow that in general promises should be kept, pain avoided, life preserved. If I have promised to do something, although in unusual circumstances I may meet your claim that I ought, therefore, to do it with a 'but' – 'but I have got bronchitis and the doctor has told me to keep indoors' – I cannot brush off your claim with a 'so what?'. Not to recognise that a promise creates a *prima facie* obligation, or that a course of action might imperil life, is to put oneself outside the pale of moral discourse. Although no claim is incontrovertible, many are not to be overlooked. I can without inconsistency commend an action which will endanger life, cause pain, contravene a previous commitment, but if I do so, I need to explain myself, either in view of wholly exceptional circumstances or else by articulating some special moral view. Practical reasoning is not closed against the possibility of new situations or new insights, but the opinions of the many and the wise do to a large extent converge about what are the relevant considerations in the ordinary run of cases.

There are many strands of argument within practical reasoning. The most primitive is the immediate decision what to do, to fight or flee, to eat or to court a mate. Much human reasoning is on this level, but we criticize the man who pursues only *to paron hedu* immediate pleasure, and responds only to immediate threats, and think it more reasonable to be prudent, taking account of long-term future interests, and not only those close at hand. Reason, we hold, is not confined to present concerns, but to future ones too. But prudence alone is not enough. It is incoherent to have regard only to the future and not to the past, and may be counter-productive to consider only myself, and not other people. Each of these points may be made with the aid of the Theory of Games, the Battle of the Sexes showing how I lay myself open to manipulation unless I take account of the past as well as the future, and the Prisoners' Dilemma showing how we fare worse if we are selfish than if we consider others' interests as well as our own.

Practical reasoning is messy. It is easy to get it wrong. If we do, we may get away with it. Although wrong decisions can lead to disaster, there is no immediate, irresistible sanction against being unreasonable. The sanction is not the breakdown of communication, but, in the absence of dire consequences, just simply a failure to be sensible, accompanied by a loss of respect in the eyes of others, especially when some moral obligation has been disregarded.

§9   Empathy and Other Minds

The two-sidedness of practical reasoning gives a key to our knowledge of other minds and our understanding of the humanities. Besides making up my mind about what I shall do, I can consider what I should do if circumstances were different; and although in the present circumstances I must over-ride and reject some considerations in accepting and acting on others, I can fully appreciate how I might in other circumstances act on them, and so I can appreciate also how you in your circumstances might act on them. Because I know what I shall do in the actual situation, I can know what I should do in hypothetical situations, and so understand what I might do if I were you. Empathy is possible because I experience in my own deliberation the conflict of argument and feel the force of factors inclining me to act in various ways. I never have murdered any one, but I have been tempted, and so can understand the minds of those who have found the temptation irresistible. Equally I can enter into the minds of historical agents or those portrayed in literature, and although sometimes their reasoning and reactions will be entirely opaque to me, often there will be enough resemblance between their situation and my actual or possible ones for their response to be one I can see the rationality of. I do not have to suppose, counterfactually and sometimes implausibly, that I *would* in the event respond in the same way, but only that I *might* – only that there would be some reasons for so acting, in the absence of weightier considerations against. And that supposition is one it is much easier to make. I can understand what makes other people tick because of the many-sidedness of what goes on in making up my own mind. The messiness of practical reasoning, and the many decisions it partially leads me to take, gives me a width of understanding I could never otherwise obtain, and a partial *entrée* into the minds of all sorts and conditions of men far beyond my actual ken.

Anger, fear, resentment, spite, greed, jealousy, gratitude, pity, love, awe, exaltation and joy commonly issue in actions and activities. Feelings are not, as too many philosophers have supposed, bare physical sensations, but are for the most part to be described in terms of what we want to do or would like to do. They are, largely, incipient actions, frustrated actions, or failed actions, and the concomitants of these. And therefore our common rationality as agents gives further and detailed support for our belief that other men are of like passions with ourselves. I think that you feel angry because I know that I would be inclined to act angrily in like circumstances; and your subsequent actions constitute a further check on whether I am right or wrong. Thus, although it is neither logically impossible nor even emotionally impossible for you to overlook the slight, or alternatively to suppress the anger which you feel, nevertheless there are always cross-connections among different states of mind and between them and actions to make it unreasonable to deny that we have some sort of fellow feeling or empathy with others, or to suppose we might be always wrong in all our ascriptions of feeling to others. *Willst du die andem verstehen*, said Schiller, *blick in dein eigenes Herz*. If you want to understand others look into your own heart. *Gnothi seauton*, Know yourself; and then you will have insight into the minds of others too.

Insight, in popular, unprofessional philosophy, is important. Historians will say of a colleague that he has his facts right and his arguments are impervious to objection and his conclusions not demonstrably wrong, yet somehow he has not got the "feel" of his period, he has not worked his way "into" it. Thus, a reviewer can write that the author

> has the rare quality of entering into the minds of those he is studying and seeing things from their point of view; the result in this case is that perhaps for the first time it is possible to understand the Aztecs and sympathize with them in their painful predicament...The book is one of the best ever written about the Aztecs: his portrait of their society is a triumph of scholarship, understanding and literary skill. [18]

On which Alan Richardson comments, "Each of these three words surely represents an essential feature of the historian's craft." [19] In the same way, literary critics will allow that a student knows all his texts, is well acquainted with the historical background, is able to manipulate parallel passages, can explain all the allusions, can comment on linguistic points, and has intelligent views on the cruces, and yet has not "got inside" his author. Scientists, it is alleged, are often insensitive: nobody denies that they are very clever men, but they lack sympathy; they are well able to conduct difficult calculations and reach right conclusions about means to given ends, but they are unable to put themselves in other men's shoes, or sense what the human reaction will be. It is claimed for the humanities that they educate men's perceptiveness, and make them more sensitive in their dealings with other men. Education apart, different men are differently endowed with this faculty: some of our friends are, and always have been, very able; they will go far; but not all of them will notice much as they go: others are more *simpatico*; they may be less clear-headed, unable to put out arguments clearly, and having a less good memory for facts and a less adequate command of argument, but nonetheless more percipient in what they say.

The untutored view is clear. Men have, besides the ability to draw conclusions from premises and to learn from experience, another, finer faculty, which enables them to "get inside". It is different from inductive reasoning both subjectively and in its operations. Subjectively, it feels different: the persons in question appear transparent, not opaque; one is unable to give as good reasons as one feels – one just knows; it is knowledge by acquaintance, not knowledge by description or by argumentation. It operates differently from induction: it does not generalise; we are quite clear that to the particular problem in mind a certain solution is the right one, but are not very ready to extend the solution to other cases; the role of evidence is unclear; we do not apply any canons of inductive inference; there seem to be no criteria of irrelevance; and often, very often, it operates when we should be unable to make any inductive inferences at all. It is as though each one of us had within him a well of singular hypotheticals, on which he was at times able to draw, and provide himself with immediate and complete solutions to certain of his problems. In ourselves we can recognise its happening: among others we can pick out those who are particularly gifted with it, those who have noses for human affairs. Sometimes it is insight into human character – Humane Insight, we might call it. Sometimes it is the ability to discard irrelevant, and fasten on important premises for an argument: this makes the good civil servant, the good historian. We value it so highly that we are constantly coining new names for it: sense, sensibility, sensitivity, being percipient, being perceptive, intuition, insight, sympathy, empathy, understanding, *verstehen*, being *en rapport*, being *simpatico*, and being able to put oneself inside other people's skins, all have been used at times to refer to it. Nor do we require it only of those engaged in practical affairs. Novelists, obviously, need to have it, and even philosophers are assessed according to, among other things, their understanding of human nature. We fault Hobbes' political doctrines because his view of man is not true to human nature. And we say this not after having examined all men and found them different from what Hobbes portrays, but merely from knowing a few men and knowing from within ourselves what it is to be a man, and what are the loyalties which can win men and move human beings to action. The attraction of Freud's doctrines lies not in the empirical evidence for them, which is often slender, but in their innate plausibility, which once Freud has expounded it, carries conviction because it corresponds to something in our own minds. We can see that this is how a man might respond, because it is how we might respond, even though we may have never responded in the fashion described, nor ever been in a situation comparable to that described. Freud's views have been accepted because he has been able to strike chords in the hearts of his readers, not because he has been able to furnish adequate statistics to establish his case. Where he is recounting some particular case or describing some typical situation, his reconstructions of the workings of the unconscious mind ring true: but where he or his disciples start trying to establish his conclusions in the way in which a medical scientist would establish his, our incredulity sets in. His conclusions are not inductive conclusions, they are not based on the evidence of our senses, nor built up from repeated, but opaque, conjunctions of observed fact. Rather, they are new ways of seeing human motives and construing human behaviour, new insights brought up from the depths of Freud's own mind. Little purpose is served by trying to assimilate this to ordinary inductive generalisation or by construing the actual experience Freud lived through himself and the hypothetical experience he could imagine for others, as a basis for an inductive generalisation in the way that our sense-experience is.

For the present we note that there seem to be inferential skills manifested by the historian and the literary critic, which do not fit into the account of inductive and Scientific argument, but can be seen as stemming naturally from the two-sidedness of practical argument. [20]

§10 Reason

Reason is very different from what many philosophers have taken it to be. Even if we start with their views, taking reason to be analytic deductive inference, we are led to a less restricted view of reason, as being typically two-sided, a dialogue between different people, conversing because they have some common objectives. Even deductive reason is not entirely analytic; and inductive arguments, and those of practical reasoning and its progeny, evidently lead to conclusions which were not implicit in the premises.

Looking back, we can now see the philosophers' ideal of valid inference as a special case, in which dialogue has been collapsed into a monologue, where the only common objective the two parties need share is to be understood. Whereas ordinarily, as I ratiocinate, my friend (or my *alter ego*), buts in with objections and counter-considerations, in an analytic deductive argument there is no room for interposing any 'but's. Any attempt to gainsay anything established by an analytic deductive argument ends in self-contradiction. As I argue, I do not need to pause, to hear objections raised, because no objection can be consistently – and therefore intelligibly – raised. Anyone who attempts to controvert what I say is contradicting himself, and can be ruled out of court as not saying anything meaningful. Monologous arguments are sometimes appropriate; in writing a book, for example, where the reader is, of necessity, in no position to interrupt. And monologous arguments are often attractive, since human nature is naturally inclined to brook no opposition. An "anti-deductivist" theme follows. Although deductive reason is the paradigm for coerciveness, it is not paradigmatic in other respects. We cannot adequately represent non-deductive inferences as deductions from additional major premises. Admittedly, sometimes, when a particular inference is in dispute, it may be helpful to articulate the rule of inference as a universal proposition, and if it proves acceptable, then the conclusion will follow deductively from it. But even in deductive logic, not all inferences can be so represented, as Lewis Carroll showed. Often also, it is unhelpful to pose the question in terms of the truth or falsity of a proposition – it makes less sense to ask whether Peano's Fifth Postulate is true than to ask whether the Principle of Recursive Reasoning is cogent. Moreover, there are many inferences in the humanities which are too particular to lend themselves to being formulated as a major premise, even though they are universalisable in a more flexible way. Reasoning is typically an open-textured dialogue to which the respondent can contribute; and the sort of contribution expected determines the structure of the dialogue. Therefore it cannot be adequately reconstructed into a deductive argument, which, since it is monologous, leaves no place for the respondent to join in. The solitary self-sufficiency of the deductivist thinker is purchased at the price of solipsistic vacuity. Reasoning is risky, and we do well to have candid friends who dare tell us that we are mistaken.

Most argument, though open-textured, is structured. Argument is fruitless, if we disagree about everything. Not only do we have to start from somewhere, but we have to have some aims and assumptions in common. Whereas simple deduction is subject to the one condition of being intelligible, usually we share a desire to know the truth, and often an aspiration to understand. Different aims and assumptions indicate different sorts of dialogue. Often in the course of one argument, we need for a time to confine the discussion to a particular type of dialogue for the sake of clarity, or in order to reach a resolution of a particular argument. It is good to be agreed about the facts of the case before embarking on interpreting them. We may need to work out the consequences of a particular hypothesis in order to test whether it is consistent with observations or falsified by them. We limit our shared commitment for the time being, in the hope of achieving some measure of agreement, before going on to more contentious matters. A serious argument is often composed of several sub-dialogues, each with its own standards of relevance and cogency. Unless we distinguish the way the different limbs are articulated into a coherent whole, we blunder, applying to one part canons only appropriate to another.

The sceptically minded may still be unwilling to go along with this account of reason. Thrasymachus may be forced to concede the cogency of simple deductive arguments, but can well refuse the invitation to enter into other men's minds. What are the sanctions against unreasonableness? They differ. I can resist the full force of Gödel's theorem by distinguishing the unprovability-within-the-system of the Gödelian formula of which there is a formally valid proof, from its being true, of which there is no formally valid proof, since there is no formal definition of the term 'true'. This, which can be seen as a consequence of Gödel's theorem, was established independently by Tarski. [21] He proved that the concept of truth cannot be formalised in any adequate formal system, because if we add to such a system a term representing what we mean by our ordinary word 'true', we shall be led – again by an Epimenides argument – to a contradiction. [22] So, when you point out the implication of denying that the Gödelian formula is true, I fail to follow your reasoning, professing not to know what 'true' means. And that is the sanction. I have divested myself of knowing what 'true' means: I have deprived myself of the concept of truth. Although syntactically I am in the clear – I have not broken any of the rules of the communication exercise – semantically I am self-mutilated – I am no longer a man of truth. The same sanction was invoked against a refusal to move from *Sorites* Arithmetic to the Rule of Recursion. In these two cases there are further, arcane sanctions. I can, without self-contradiction, deny the truth of the Gödelian formula: there is no inconsistency between it and the axioms of elementary number theory. It follows that there is a model of the negation of the Gödelian formula and the axioms of elementary number theory. It is a weird model, but it cannot be faulted on formal grounds. If I refuse to acknowledge the truth of the Gödelian formula, I cannot exclude weird models, and so cannot specify that the numbers I am talking about are the same as the numbers that you, and everybody else, are talking about. Somewhat similarly, if I refuse to accept the Rule of Recursion, although I can specify each individual number separately, I cannot talk about them all collectively. I could be guilty of "ω-inconsistency", alleging that although each natural number possessed some property, some did not. The sceptic, who will not accept the truth of the Gödelian formula, or the validity of the Rule of

Recursion, is not guilty of any straightforward inconsistency, but will be talking at cross purposes when he argues with those who have a firm grasp on the nature of the natural numbers.

Inductive sceptics have difficulty in living down to their unbeliefs. If I refuse to anticipate future events, I am in for nasty surprises. Even if my animal instincts enable me to avoid disaster in the particular situations I find myself in, I am handicapped, if I cannot generalise the better to communicate to others and to remember myself what experience has taught me. And the price of not inferring to the best explanation is not to have the best explanation. I can live without understanding why things happen as they do. It is my choice. But if I choose to do so, I am the sufferer.

In practical affairs, again, I shall be the chief sufferer if I do not use reason, though others may suffer too as a result of my imprudence, lack of consideration, lack of public spirit, or lack of commitment to anything outside myself. Traditionally, philosophers have drawn a sharp distinction between counsels of prudence and precepts of morality, and have thought the former unproblematic, while seeing great difficulties in showing why we ought to be moral. There is indeed a distinction, but it is not as clear-cut as has been made out, and the sanctions are not as sharply separated as commonly supposed. Others, as well as I, may suffer if I am imprudent, and I, as well as others, if I am selfish. In part it is a question of identity and identification: if I have no consideration for others, I debar myself from meaningful use of the first person plural, and isolate myself into the logical loneliness of Plato's autistic autocrat; and if I am entirely insensitive to any obligation to any objective value, I cannot ground my agency, my actions, or my achievements in anything of greater worth than my own fleeting preferences. I do not have to enter into your concerns – indeed, I economize on emotional drain, if I do not waste sympathy on you in your misfortunes; in some cases the balance of advantage may go the other way – I shall gain more from your co-operation than I shall lose in bearing your troubles, but, whatever the balance of crude advantage, I lose out in knowledge and understanding if I make myself ignorant of the working of other men's minds, reducing my range of sensibility and diminishing myself.

Sanctions not only underwrite cogency, but impose a strategy. If you are obdurate, and do not feel the force of my argument, I must manoeuvre you into a tight corner, where you will pay the penalty for your obduracy. Hence the pressure to formulate; hence also the importance of chains of derivations. If I give a derivation, I can challenge you to say where I am wrong, and you have to specify, and I can then concentrate on that point. Especially in mathematical argument, I can often show you to be inconsistent. In physics, I may show that you would be having to deny some important symmetry, or other rational requirement. Often in physics and in other sciences too the cost of resisting reason is a loss of understanding. If I persist in not accepting evolution, I shall not attain the wide-ranging perspective on how living creatures are related to one another that the theory of evolution offers, and shall not understand how different species came to be. At a more mundane level unpleasant consequences flow from not believing Scientific laws and maxims of common sense; and in a different way from not cooperating with others, and treating them well. The experience of living, unshielded by parental care, in a small community is often the means of instilling common sense and common decency by exposing the foolish and fickle to inescapable peer-group pressure.

Normative reason has an edge. It arises from our being able to think wrong. I can think wrong, you can think wrong, and we want to convict others' wrong thinking of error. Equally, in view of similar desires on the part of others, we seek to make our own arguments invincible. In either case the strategy is polemical. Each side seeks to pin down its opponents' arguments, and search for weak spots, where a decisive victory can be gained; and, anticipating similar moves on the part of the other side, tries to formulate defences against hostile probing. But some reasonings have no sanctions and no strategy for securing agreement. I give my reasons for having acted as I did, and you may be able to share them, and accept them as telling, but if you do not feel their force, and do not accept them, your only loss is not to be able to understand why I acted as I did.

Sanctions

1. Analytic statements and First-order Logic: Failure to communicate.

2. Gödel's theorem: No grasp of the concept of truth; inadequate grasp of the concept of number.

3. Rule of Recursion: No grasp of the concept of truth; ω-inconsistency; inadequate grasp of the concept of number.

4. Inference to the Next Case: Nasty surprises.

5. Inference to a Generalisation: No General Grasp.

6. Inference to Best Explanation: Inexplicability.

7. Practical Reason: No common sense.

8. Humane Insight: No understanding of fellow men.

The development of the idea of normative reason, starting from the minimal requirement for our communications to be intelligible, and moving through successive stages to a wide-ranging ability to enter into the minds of others, and to argue about what we, individually or collectively, ought to do, reveals a characteristic universality. If an argument is cogent on one occasion, it will be cogent on others too. Hence the move from *Sorites* arguments to the Principle of Recursive Reasoning, and from Induction to the Next Case to Inductive Generalisation. In practical reasoning, especially in moral, political and judicial argument, a similar move is made, giving rise to a principle of universalisability. Kant formulated it as the command: "Act only on that maxim through which you can at the same time will that it should become a universal law". [23] But although often it is legitimate to argue that what is sauce for the goose is sauce for the gander, we have to recognise, at least where human beings are concerned, that one man's meat is another man's poison. Some principle of universalisability remains valid, but it needs to be a more flexible one, not only in moral discourse, but in disentangling historical causes, and evaluating arguments in the humanities generally. Instead of positing some principle such that all cases falling under it are taken to be the same, we should require only that if some apparently similar case is taken to be different, a reason for the difference should be forthcoming.

If reason is universal, it can reason about reason itself, and from self-referential reasoning, it emerges that it is not possible to set bounds to reason, and in particular that metaphysical argument is not beyond the bounds of reason. Deductive argument gives rise not only to recursive arguments, but to Gödel's self-referential theorems; and inductive arguments merge into Inference to the Best Explanation, sometimes invoking entities beyond the bounds of possible experience. Practical reasoning leads on to empathy and moral argument, and they in turn lead to the humane insight of the humanities, and varieties of political, legal and judicial argument. Gödel's theorem shows that even with deductive argument, we cannot formalise completely; however far we formalise our rules of inference, there will still be some inference which is clearly valid but does not fall under any of the rules thus far formulated. Even deductive argument is fuzzy-edged, and the transition from one type of inductive inference to another shows that the same holds good for inductive arguments about matters of fact and their explanation; and counts against the contention that reason cannot lead on to the various styles of practical argument.

We can go further. The simple argument of the Logical Positivists against metaphysics does not work. If metaphysics could be ruled out by the Verification Principle, so too would the Verification Principle itself. Similarly, any claim to set a boundary to reason can be challenged. Clearly, if the bounds of reason are so tightly drawn as to exclude philosophical argument, the claim will exclude any possible justification of itself. But more generally, in order to determine the boundary exactly, it will need to specify what lies beyond it and is to be excluded, as well as what lies within it and is to be included. And if the boundary is really a boundary reason cannot overstep, reason will be precluded from stepping over it, to specify precisely what is to be excluded. Reason itself, then, is unbounded. We have a negative, self-referential argument, analogous to the negative, self-referential mathematical argument underlying Gödel's theorem, against any claim that reason can be corralled within any antecedently set limit, and are led to a crucial conclusion. Reason is not "thin".

Theses about Reason

1. Not all a priori arguments are analytic. For example, in mathematics, arguments by recursion: (point made by Poincaré): it is not straight inconsistent (though it is ω-inconsistent) to hold for *each* natural number, *n;*

that $F(n)$, but to deny that $(\forall n)F(n)$,

that is, for *all* natural numbers $F(n)$.

2. Not all deductive arguments are rule-bound, that is governed by some antecedently specified rule of inference. For example, Gödel's theorem shows that the Gödelian sentence of a formal system is true, though this cannot be proved in the system. We need to know how to argue, not just know that certain rules of inference are allowed. [Achilles and the Tortoise]

3. Not all sound arguments are deductive. For example, inductive arguments.

4. Not all inductive arguments are of the same type. For example, some inductive arguments argue from particular propositions as premises to another particular proposition as conclusion (the sun has risen every day hitherto, so it will rise again tomorrow), others argue from particular propositions as premises to a general proposition as conclusion (the sun has risen every day hitherto, so it rises every day).

5. Not all arguments about the way the world is are simple inductive arguments. For example, many Scientific arguments go from observational premises to theories about entities that cannot themselves be observed.

6. Not all arguments go from premises to conclusions of the same logical type. For example those instanced in (4) and (5).

7. Some arguments go from factual premises to evaluative conclusions. For example, moral arguments; also intellectual arguments about the acceptability of conclusions of arguments.

8. Most arguments, apart from deductive arguments, are two-sided. The addition of further information may weaken, not strengthen, a conclusion; they are not monotonic, but are a matter of argument and counter-argument, objection and rebuttal. For example, induction (even though I have seen 257 swans, and they are all white, and it was reasonable on that evidence to infer that all swans are white, if I now come across a black swan, I can no longer assert that all swans are white). The point has long been known in practical reasoning, where the very word 'deliberation' suggests weighing considerations pro against considerations con.

9. *Tertium non datur* (the principle of excluded middle).

10. Reason is creative.

11. Cumulative arguments.

12. No venture, no win (nothing ventured, nothing gained).

13. Fallibility.

14. Self-reflective: metaphysical arguments often make use of this; as when we refute a metaphysical argument (e.g. the Verification Principle), on the grounds that it is sawing off the branch it is sitting on.

Endnotes

1. The use of the word 'valid' has caused difficulties. It is sometimes used as a general term of appraisal, at other times as applying only to deductive arguments. See J.O.Urmson, "Some Questions Concerning Validity", *Revue Internationale de Philosophie*, 25, 1953, pp.217-229; reprinted in R.G.Swinburne, *The Justification of Induction*, Oxford, 1974, pp.74-84. In this chapter it will be used only of deductive arguments, which are either valid or invalid, with no further value in between. Inductive and other arguments will be assessed as being of greater or less cogency. It follows that an "invalid" (i.e. not *deductively* valid) argument may nevertheless be extremely cogent.

2. The word 'logic', like the word 'valid' is the cause of much confusion. Often it is taken to mean 'deductive logic', but philosophers used to talk of 'inductive logic', without its being a contradiction in terms. Scientists sometimes speak of the logic of an experiment, and historians of the logic of a situation, while feminists are furious when men say that women are emotional rather than logical. Confusion is best avoided by always asking what the words 'logic' and 'logical' are being contrasted with.

3. The older logicians wrote *not both (p and not q)*, as $-(p \& -q)$, but it is better to have a distinct sign for negation, $\neg$, which will not be mistaken for a dash. (Also note that instead of the simple ampersand, $\&$, logicians often now prefer to use a wedge-shaped symbol.)

4. Indeed, we can generate all the truth functions from a single binary 'neither . . nor . . ', $\downarrow$, joint denial, where $p \downarrow q$ is true just in case both p and q are false, or from $|$, alternative denial, which is false just in case both p and q are true. Whitehead and Russell took $p \vee q$ as basic, defining $p \rightarrow q$ as $\neg p \vee q$, and $p \& q \neg(\neg p \vee \neg q)$.

5. Lewis Carroll, "What the Tortoise Said to Achilles", *Mind*, 10, 1895, pp.278-290; reprinted in *The Works of Lewis Caroll*, ed. R.L.Green, London, 1965, pp.1049-51. See also, Gilbert Ryle, *The Concept of Mind*, London, 1949, ch.2, pp.25-61; or "Knowing How and Knowing That", *Proceedings of the Aristotelian Society*, 48, 1945-1946, pp.1-16; and Kant, *Critique of Pure Reason*, tr. Kemp Smith, A132-3/b171-2.

6. K. Gödel, "Über Formal Unentscheidbare Sätze der *Principia Mathematica* und verwandter Systeme", Part I, *Monatschefte für Mathematik und Physik*, Vol. XXXVIII (1931), 173-198. Reprinted with English translation in *Kurt Gödel: Collected Works*, vol.1, Oxford University Press, 1986, pp.144-195. There is also an English translation in Kurt Gödel, *Lectures at Institute of Advanced Study*, Princeton, N.J., 1934; and one by B.Meltzer, Edinburgh, 1962 (but see review in *Journal of Symbolic Logic*, 30, 1965, pp.357-359). In this paper Gödel proved two theorems. It is his first theorem that is here discussed.

7. H.Poincaré, *Science and Hypothesis*, pbk ed., New York, 1952, §§IV-VII, pp.8-13.

8. Although it may be advantageous, for the purposes of logistic analysis, to articulate an axiom, which can be added or not added to an axiomatic system, we gain a deeper understanding of the principle involved, if we view it as a rule of inference. If we do regard it as an axiom, it is evidently a synthetic *a priori* one, stating some fact about a strange universe of enormous size: we wonder not only whether it is true or false, but also how we could ever come to know its truth or falsity. A rule arising from a dialogue between two truth-seekers is much easier to understand and to justify.

9. See J.R.Lucas, "On Not Worshipping Facts", *The Philosophical Quarterly*, 8, April, 1958, pp.143-156.

10. Some care is needed in expressing this inference, in order not to make it a tautology – days are often defined in terms of the sun's rising, so that tomorrow would not be tomorrow unless the sun rose.

11. Bertrand Russell, *The Problems of Philosophy*, 2nd ed., Oxford, 1946, p. 63; reprinted in R.W.Swinburne, ed., *The Justification of Induction*, Oxford, 1974, p.21.

12. See previous section.

13. K.R. Popper, *Logik der Forschung*, Vienna, 1934/5, tr. as *The Logic of Scientific Discovery*, London, 1959.

14. J.J.C. Smart, *Philosophy and Scientific Realism*, London, 1963, p.133. Smart uses italicised but otherwise grammatically correct present tenses; for reading aloud the grammatically incorrect use of the infinitive is clearer. See, more fully, Nicholas Rescher, "On the Logic of Chronological Propositions", *Mind*, 75, 1966, pp.75-76.

15. Kant, *Groundwork of the Metaphysic of Morals*, tr. H.J.Paton, as *The Moral Law*, London, 1947, p.88.

16. See, more fully, J.R.Lucas, "The Lesbian Rule", *Philosophy*, 30, 1956, pp.195-213.

17. See further, J.Raz, *Practical Reasons and Norms*, London, 1975, and Princeton, 1990, ch.1, §1.1, pp.25-35; see also S.E.Toulmin, *The Uses of Argument*, Cambridge, 1958, ch.3, pp.97-102.

18. *Times Literary Supplement,* Nov. 10, 1961 (No.3, 115), p. 800; the work under review is Jaques Soustelle, *The Daily Life of the Aztecs*, Eng. trans. by P.O'Brian, London, 1961.

19. A.R.Richardson, *History Sacred and Profane*, London, 1964, pp. 158-159.

20. J.H.Newman, *The Grammar of Assent*, London, 1870, ch. VIII, S 2, section 2, pp. 296-7; W.H.Walsh, *Introduction to the Philosophy of History*, London, 1951; and W.H.Dray, *Laws and Explanation in History*, Oxford, 1957. For further consideration of the role of empathy in history see C. Portal, ed. *The History Curriculum for Teachers*, Falmer Press, 1987.

21. A.Tarski, "The Concept of Truth Formalized Languages" tr. J.H.Woodger in *Logic, Semantics, Metamathematics*, Oxford, 1956, Ch.VIII.

22. See above, §1.

23. *The Groundwork of the Metaphysic of Morals*, 421/52, tr. H.J.Paton, *The Moral Law*, London, nd.(1948), p.88.

Acknowledgement

This book chapter has been printed by permission of J. R. Lucas; it has been adapted by him from (chapter two of) his book **Reason and Reality** (forthcoming in 2009 from Ria University Press). The copyright and intellectual property rights of this chapter belong to J. R. Lucas.

CHAPTER SEVEN

The Basic Ideas Of Conformal Cyclic Cosmology[1]

Roger Penrose

1. Conformal Space-Time Geometry

In this article, I outline a new cosmological proposal—
conformal cyclic cosmology, or CCC—according which the
universe undergoes repeated cycles[2] of expansion, that I refer
to as *aeons*, each starting from its own "big bang" and finally
coming to a stage of accelerated expansion which continues
indefinitely (which would be for an infinite time, according
to how an accurate physical clock would measure time), in
close accordance with current observations[3] of our own aeon.
There is no stage of contraction (to a "big crunch") in this
model. Instead, each aeon of the universe, in a sense
"forgets" how big it is, both at its big bang and in its very
remote future where it becomes physically identical with the
big bang of the next aeon, despite there being an infinite
scale change involved, on passing from one aeon to the next.

To put this in more mathematical terms, we must consider
the geometrical structure that becomes of relevance at the
cross-over from aeon to aeon. This is what is referred to as
conformal structure rather than a metric structure. Conformal
structure may be viewed as family of metrics that are
equivalent to one another via a *scale change*, which may vary
from place to place. Thus, in conformal space-time geometry,
we do not have any particular metric g_{ab}, but an equivalence
class of metrics where the metrics $\hat{g}_{ab}$ and g_{ab} are considered
to be equivalent if there is a smooth positive scalar field Ω
for which $\hat{g}_{ab}=\Omega^2 g_{ab}$. A more directly physical way of putting
this is to say that, for a space-time, its conformal structure is
simply its *light-cone* structure. It will be appropriate to

incorporate the distinction between *future* and *past* light cones into the notion of "conformal structure", this structure being then equivalent to the space-time's *causal structure*.[4] We may note that most of the information in the metric is actually in the conformal (or causal) structure, because only 1 out of the 10 components (per point) of the metric is needed to fix the scale, whereas the remaining 9 (as ratios) are what fix the locations of the light cones. It is this conformal structure which is to remain smooth at the cross-over from one aeon to the next.

The reader may wonder why one should consider it necessary to adopt this strange-seeming view of the history of the universe. There are, in fact, several different ingredients which point the way to such a proposal, various aspects of which are implicit in the nature of the cosmic microwave background (CMB). It has become customary—essentially part of what is now referred to as the "concordance" picture of cosmology—to try to explain these in terms of the concept of "inflation", whereby a substantial period of exponential expansion it taken to be part of the universe's extremely early history.[5] However, inflation does not remotely come to terms with what I would regard as the most fundamental conundrum of all, concerning the extraordinarily special nature of the Big Bang,[6] and it is this that provided the initial motivation underlying CCC.

2. The Basic Conundrum

Proposals for describing the initial state of the universe hardly ever address a certain fundamental issue—yet this is an issue whose significance is, in a certain sense, obvious. This arises from one of the most fundamental principles of physics: the *Second Law of thermodynamics*. According to the Second Law, roughly speaking, the entropy of the universe increases with time, where the term "entropy" refers to an appropriate measure of *disorder* or lack of "specialness" of the state of the universe. Since the entropy increases in the future direction of time, it must decrease in the past time-direction. Accordingly, the initial state of the universe must be the most special of all, so any proposal for the actual nature of this initial state must account for its extreme specialness. Proposals have been put forward from time to time (such as in various forms of inflationary cosmology and the previously popular "chaotic cosmology"[7]) in which it is suggested that the initial state of the universe ought to have been in some sense "random", and various physical processes are invoked in order to provide mechanisms whereby the universe might be driven into the special state in which it appears actually to have been in, at slightly later stages. But "random" means "non-special" in the extreme; hence the conundrum just referred to.

Sometimes theorists have tried to find an explanation via the fact that the early universe was very "small", this smallness perhaps allowing only a tiny number of alternative initial states, or perhaps they try to take refuge in the *anthropic principle*, which would be a selection principle in favour of certain special initial states that allow the eventual evolution of intelligent life. Neither of these suggested explanations gets close to resolving the issue, however.[8] It may be seen that, with time-symmetrical dynamical laws, the mere smallness of the early universe does not provide a restriction on its degrees of freedom. For we may contemplate a universe model in the final stages of collapse. It must do *something*, in accordance with its dynamical laws, and we expect it to collapse to some sort of complicated space-time singularity, a singularity encompassing as many degrees of freedom as were already present in its earlier non-singular collapsing phase. Time-reversing this situation, we see that an initial singular state could also contain as many degrees of freedom as such a collapsing one. But in our actual universe, almost all of those degrees of freedom were somehow not activated.

What about the anthropic principle? Again, this is virtually no help to us whatever in resolving our conundrum. It is normally assumed that life had to arise via complicated evolutionary processes, and these processes required particular conditions, and particular physical laws, including the Second Law. The Second Law was certainly a crucial part of evolution, in the way that our particular form of life actually came about. But the very action of this Second Law tells us that however special the universe may be now, with life existing in it now, it must have been far *more* special at an earlier stage in which life was not present. From the purely anthropic point of view, this earlier far more special phase was not needed; it would have been much more likely that our present "improbable" stage came about simply by chance, rather than coming about via an earlier even more improbable stage. When the Second Law is a crucial component, there is always a far more probable set of initial conditions that would lead to this same state of affairs, namely one in which the Second Law was *violated* prior to the situation now!

As another aspect of this same issue, we may think of the vastness of our actual universe, most of which had no actual bearing on our existence. Though very special initial conditions were indeed required for our existence in our particular spatial location, we did not actually need these same special conditions at distant places in the universe. Yet as we look out at the universe, we see the same kind of conditions, acting according to the same Second Law of thermodynamics, no matter how far out we look. If we take the view that the Second Law was introduced in our vicinity merely for our own benefit, then we are left with no explanation for the extravagance of this same Second Law having to be invoked uniformly throughout the universe, as it appears to be as far as our powerful instruments are able to probe.

3. The Enormity of the Specialness

In order to stress the extraordinary scale of this problem, and the intrinsic implausibility of explanations of this kind, it is helpful to enter a little more precisely into the definition of entropy, and to estimate the entropy magnitudes that we have to contend with. Boltzmann provided us with a beautiful formula for the entropy S of a system:

$$S = k \log V.$$

Here k is Boltzmann's constant and V is the volume of a certain region in the total *phase space*[9] P of the system under consideration. We are taking P to be "coarse-grained" into sub-regions, each sub-region representing states that are deemed to be indistinguishable with regard to any reasonable macroscopic parameter. (There is clearly an element of arbitrariness or subjectivity, here, as to which parameters are to be regarded as macroscopically discernible and which are deemed to be effectively "unmeasurable". In practice, there is a considerable robustness with regard to this arbitrariness, and it is reasonable to disregard this issue in the present discussion.) Any particular state of the system under consideration will be specified by some point x of P, and the quantity V is then the volume of the particular sub-region of P which contains x.

With regard to future time-evolution of the system, the Second Law can be understood as the fact that, as the system evolves, the point x moves within P so that with overwhelming probability it enters sub-regions of successively larger and larger volume V. This arises from the fact that, in practice, the sub-regions differ stupendously in size. The logarithm in Boltzmann's formula helps here (as does the smallness of k, in ordinary units), because there need only be a modest increase in S when x moves from one sub-region into a neighbouring one of stupendously larger volume. But this is only the easy half of our understanding of the Second Law. The difficult half is to understand why, when we *reverse time*, x enters successively *tinier* sub-regions of P. It does this because it has ultimately to reach the exceptionally tiny region B which represents the Big Bang itself. The difficult half of the Second Law involves an understanding of why the universe had to start off in such an extraordinarily special state. And to understand how special the Big Bang actually was, we need to compare the volume of B with that of the entire phase space P.

One point of concern is the fact that the entire volume might be infinite, as it certainly would be in the case of a spatially infinite universe. This issue, while of relevance, is not of major importance for our considerations here. There is also the issue of how we get a finite phase-space volume when some of the parameters would be describing continuous fields. I shall evade this latter issue by assuming that it is dealt with by quantum mechanics, where for a finite universe of bounded energy content we may assume only finitely many quantum states.

To deal with a spatially infinite universe, I shall assume that we need consider only, say, that *co-moving*[10] portion of the universe that intersects our past light cone. This contains something of the order of 10^{80} baryons. To obtain a lower bound for the volume of P, for this situation, we can consider the entropy that arises when this number of baryons is collapsed into a black hole. For this, we use the Bekenstein–Hawking entropy formula[11] $S_{BH}=8\pi^2kGm^2/hc$ for a spherical black hole of mass m and find a value of the order of 10^{123}. If we envisage the *dark matter* also being absorbed into this black hole, we would get a considerably larger entropy (and, for a continually expanding universe, we must contemplate possibly even larger entropy values that might be attained in the very remote future), but this value represents a usable lower bound. Recalling the logarithm in Boltzmann's formula (this being a natural logarithm, but that is of no concern), we find that the volume of P is greater than that of B by a factor that exceeds

$$10^{10^{123}}.$$

This gives us some idea of the enormity of the precision in the Big Bang!

4. The Geometric Nature of the Specialness

A seeming paradox arises from the fact that our best evidence for the very existence of the Big Bang arises from observations of the microwave background radiation—frequently referred to as the "flash of the Big Bang", greatly cooled down to its present value of ~2.7K. The intensity of this radiation, as a function of frequency, matches the Planck radiation formula extraordinarily closely, giving us impressive evidence of an early universe state with matter in *thermal equilibrium*. But thermal equilibrium is represented, in phase space P, as the coarse-graining sub-region of largest volume (so large that it normally exceeds all others put together). This corresponds to *maximum* entropy, so we reasonably ask: how can this be consistent with the Second Law, according to which the universe started with a very *tiny* entropy?

The answer[12] lies in the fact that the high entropy of the microwave background refers only to the *matter* content of the universe and not to the *gravitation field*, as would be encoded in its space-time geometry in accordance with Einstein's general relativity. What we find, in the early universe, is an extraordinary *uniformity*, and this can be interpreted as the gravitational degrees of freedom that are potentially available to the universe being not excited at all. As time progresses, the entropy rises as the initially uniform distribution of matter begins to clump, as the gravitational degrees of freedom begin to be taken up. This allows stars to be formed, which become much hotter than their surroundings (a thermal imbalance that all life on Earth depends upon), and finally this gravitational clumping leads to the presence of black holes (particularly the huge ones in galactic centres), which represent an enormous increase in entropy.

Although, in general, there is no clear geometric measure of the entropy in a gravitational field in general relativity, we can at least provide proposals for the *non-activation* of gravitational degrees of freedom at the Big Bang. I have referred to such a proposal as the *Weyl Curvature Hypothesis*[13] (WCH). In Einstein's theory the *Ricci curvature* R_{ab} is directly determined by the gravitational sources, via the energy-momentum tensor of matter (analogue of the charge-current vector J^a in Maxwell's electromagnetic theory) and the remaining part of the space-time Riemann curvature, namely the *Weyl curvature* C_{abcd}, describes gravitational degrees of freedom (analogue of the field tensor F_{ab} of Maxwell's theory). WCH—which is a time-*asymmetrical* hypothesis—asserts that *initial* space-time singularities must be constrained to have $C_{abcd}=0$ (in some appropriate sense), whereas final space-time singularities (as occur inside black holes) are unconstrained.

What appears to be the most satisfactory form of WCH has been studied extensively by Paul Tod[14]. This proposes that an initial space-time singularity can always be represented as a smooth *past boundary* to the *conformal geometry* of space-time. Tod's formulation of WCH is the hypothesis that we can adjoin a (past-spacelike) hypersurface boundary to space-time in which the conformal geometry can be mathematically extended smoothly through it, to the past side of this boundary. This amounts to "stretching" the metric by a conformal factor Ω which becomes *infinite* at the Big

Bang singularity, so that we get a smooth metric $\hat{g}_{ab}$ which actually extends across this boundary.

5. Conformal Cyclic Cosmology

So far, we regard the conformal "space-time" prior to the Big Bang as a mathematical fiction, introduced solely in order to formulate WCH in a mathematically neat way. However, CCC takes this mathematical fiction seriously as something *physically real*. But what "physical reality" can we consistently attach to this space-time occurring "before the Big Bang"? As a clue to this possibility, we should consider the nature of the physics that is presumed to be taking place just *after* the Big Bang. (I am ignoring the possibility of inflation here, as CCC seems to provide an alternative explanation for the main positive achievements of inflation; see §6.) As we approach the Big Bang, moving back in time, we expect to find temperatures that are increasingly great And the greater the temperature, the more irrelevant the rest-masses of the particles involved will become, so these particles are effectively massless near the Big Bang (presumably at energies that were greater than that provided by the mass of the Higgs particle). Now, massless particles (of whatever spin) satisfy conformally invariant equations.[15] I am going to suppose that the *interactions* between these massless entities are also described by conformally invariant equations. (This seems to be consistent with current understanding of particle physics.) With such conformal invariance holding in the very early universe, the universe has no way of "building a clock". So it loses track of the scaling which determines the full space-time metric, while retaining its conformal geometry.[16]

We may apply considerations of this kind also to the distant future of the universe. If we assume that in the very remote future conformally invariant equations again govern the universe's contents, then we can apply the same mathematical trick as before, but now in the reverse sense that we look for a boundary at which the conformal factor Ω becomes *zero*, rather than infinite. This amounts to using a metric, such as $\hat{g}_{ab}$ above, in which the future infinity is "squashed down" to be a finite boundary to space-time, which is conformally regular in the sense that the space-time can be mathematically extended across this future boundary as a smooth conformal manifold.[17] If we also assume that there is a positive cosmological constant present, as current observations appear to point strongly towards,[18] then we find that this future conformal boundary is *spacelike*.

There is, however, a crucial difference between the use of a conformal boundary to study the future asymptotics of a space-time and Tod's use of a conformal boundary to treat the Big Bang. For in the latter case the very validity of this trick provides a formulation of WCH, whereas in the future situation of an expanding universe with conformally invariant contents, the validity of this procedure is more-or-less *automatic*[19]. Physically, we may think that again in the very remote future, the universe "forgets" time in the sense that there is no way to build a clock with just conformally invariant material. This is related to the fact that massless particles, in relativity theory, do not experience any passage of time. We might even say that to a massless particle, "eternity is no big deal". So the future boundary, to such an entity is just like anywhere else. With conformal invariance both in the remote future and at the Big-Bang origin, we can try to argue that the two situations are *physically identical*, so the remote future of one aeon of the universe becomes the Big Bang of the next. This is the basis of CCC.

6. Physical Implications

There are certain important assumptions involved in CCC, in order that only conformally invariant entities survive to eternity. One of these is that black holes will all eventually evaporate away and disappear. This evaporation is a consequence of Stephen Hawking's quantum considerations,[19] and these are now normally accepted. There is the issue of whether black holes would actually ultimately disappear or perhaps leave some form of "remnant". I am here taking the more conventional view that they would indeed disappear in a final (cosmologically very mild) explosion. However, there is another aspect to this concerning the so-called "information paradox", and I shall need to return to this in §8.

A more immediate issue, for CCC, is how to get rid of massive fermions and massive charged particles. It is not too unconventional to assume that protons will ultimately decay, or even that there could be one variety of neutrino that is massless, but the real problem lies with electrons. A good many of them will annihilate with positively charged particles, but there will be a relatively small number of "stray" charged particles (electrons, positrons, and also perhaps protons if these do not decay) which become trapped in their ultimate cosmological event horizons, being unable to come in contact with other particles of opposite charge. There are various possible ways out of this, none of which is part of conventional particle physics. One possibility is that electric charge is not exactly conserved, so that within the span of eternity, electric charge would eventually disappear. A much more satisfying possibility, is that *rest-mass* is not, in a sense, completely constant, so that the electron's mass will eventually decay away—and, indeed, so would the rest-mass of all massive particles (e.g. neutrinos), perhaps at different rates, albeit extremely slowly, over the infinitude of eternity. This seems to me to be not at all implausible, because the very reason that rest-mass is a "good quantum number" is the result of it being a *Casimir operator* for the Poincaré group (i.e. commuting with all the generators of the group). However, when there is a cosmological constant present (as is crucial for CCC in any case), the Poincaré group is not completely correct for understanding particle physics, and we need to turn to the deSitter group for something more appropriate. The rest-mass is *not* a Casimir operator for the deSitter group, so it may not be implausible for there to be a very slow decay of rest-mass over enormous stretches of time. It should be made clear that I do not mean the decay of massive particles into massless ones, but a universal decay of rest-mass itself, which is not necessarily at the same rate for all types of particle. Accordingly, charged particles such as electrons and positrons would asymptotically[20] lose their mass, so that in the asymptotic limit of the remote future, all mass is lost, so conformal physics again becomes the relevant physics, and conformal geometry the relevant space-time geometry, just as it had been at the Big Bang.

These matters are all tied up with the issue of the strength of the gravitational interaction, which I have postponed in my discussion here. In the background of conformal geometry, the strength of gravity may be considered as being infinitely large at the Big Bang (which is, in a sense, why the gravitational degrees of freedom must initially be set to zero), and this strength gets smaller as time progresses, eventually reducing to zero at the final boundary. To express all this in a satisfactory mathematical framework for CCC, we need to reformulate general relativity in an appropriately conformally invariant way. This can indeed be done. We take advantage of the fact that the Weyl tensor $C_{abc}{}^{d}$ is conformally invariant, and provides a precise measure of the conformal curvature of space-time. We can define the gravitational "spin-2 field" $K_{abc}{}^{d}$ to be described by $C_{abc}{}^{d}$ with respect to the original space-time metric g_{ab}, but when we pass to the conformally related metric $\hat{g}_{ab}=\Omega^{2}g_{ab}$ we find that, curiously, $K_{abc}{}^{d}$ picks up a factor of Ω^{-1}, which $C_{abc}{}^{d}$ does not.[21]

This has the implication that gravitational radiation (described by $K_{abc}{}^{d}$) actually survives at the future boundary (whereas $C_{abc}{}^d$ vanishes there) and its presence shows up as a non-zero normal derivative of $C_{abc}{}^d$ at the boundary. This gives rise to primordial density and velocity fluctuations at the Big Bang. The details of all this have yet to be fully worked out, but, in principle at least, there should be clear-cut predictions which should be observable. Some of these will be discussed in the next section.

7. Observational Effects in the CMB

There are numerous observational consequences of CCC, the most immediate of which would appear to be the way in which it should lead to temperature variations in the cosmic microwave background. The observed variations are at the level of only about 1 part in 10^5. The presently conventional view of how these come about is via quantum fluctuations at the Big Bang becoming stretched out by inflation to cosmological scales. The exponential—and essentially scale-invariant—expansion, that is taken to be present in the inflationary phase, would provide a distribution of density variation, and hence temperature variation, that is scale invariant, this comparing rather favourably with detailed CMB observations over a broad spectrum of sizes. This scale invariance is taken a success of inflationary theory, whereas in CCC there is no inflation, as such. But it would appear that CCC has implications that are rather similar, in this respect, since it also involves an exponential expansion, though this occurs *before* the Big Bang in CCC, rather than afterwards.

The origin of the irregularities, according to CCC would have to be something quite different from quantum fluctuations, however. It may be that irregularities in the ultimate energy distribution in the previous aeon would simply propagate into the next aeon. But there is a particular factor which could provide a very significant and interesting input into the energy and velocity distribution of the primordial material in the subsequent aeon. This would be the effect of encounters between super-massive black holes that inhabit the centres of galaxies. These would be expected to occur at the later stages of the aeon previous to ours, and the gravitational waves emitted in such encounters could well carry away an appreciable proportion of the mass content of the black holes themselves in a relatively brief burst of gravitational energy. Owing to effects discussed at the end of §6, these bursts ought to give rise to circular rings along which the CMB temperature should appear either slightly warmer or slightly cooler than the average CMB temperature, depending mainly upon whether the source is very distant or comparatively close, as compared with the "particle horizon" of our current aeon. The effect ought to be somewhat similar to the pattern of ripples on a pond following a period of rain which had recently stopped: each raindrop produces a circular ring, but a detailed statistical analysis would be required in order to ascertain whether the resulting pattern of ripples is composed in this way. A preliminary search by Amir Hajian, under the direction of David Spergel, is inconclusive, so far, and further results are awaited.

8. The Second Law of Thermodynamics

We must return, finally, to the issue of the Second Law of thermodynamics, which initiated this entire line of thinking. Although the very structure of CCC demands a very particular structure for the big bang of every aeon, there is a conundrum still remaining. For in matching the aeon's future conformal surface, which represents what we expect to take as a very high entropy state representing the remote future of any particular aeon, with the very low entropy big-bang state of the succeeding aeon, we find the seeming contradiction that every surviving conformally invariant field must match, one side to the other. Thus, each photon goes exactly to a corresponding photon, as it traverses the 3-surface joining one aeon to the next. So also must each charged (or neutral) particle—bereft of its mass in accordance with the discussion of §6. In the case of gravitation, the degrees of freedom also match, although the gravitational information in the ultimate radiation of the previous aeon is converted into spatial conformal curvature and the density distribution and initial motions of the next aeon's new primordial (scalar) material.

This would appear to be a serious paradox, since we must suppose that, during the course of the evolution of each aeon, where the Second Law is expected to hold throughout, that the entropy should be enormously larger at the ultimate state of that aeon than it was at its big bang. The answer to this puzzle is a somewhat subtle matter. In observable physical processes, there is no violation of the Second Law, but to understand this we must return to the Boltzmann definition of entropy, as given in §3. We require knowing the volume V of the relevant coarse-graining region phase space appropriate to the state of the universe, and for that we need to know the appropriate phase space for the universe. I have already drawn the reader's attention, in §6, to the fact that, during most of the universe's evolution, by far the largest entropy increasing process is the formation of black holes. When a black hole grows through swallowing matter, the entropy continues to increase, but eventually, when the universe cools down and thins out, as it is expected eventually to do owing to its (accelerated) expansion, the temperature of the universe becomes cooler even than the tiny (Hawking) values that black holes have. At that point the hole shrinks in size as mass-energy is carried away in Hawking radiation. The process has the effect of continuing to be in accordance with the Second Law, but there is a subtlety here. According to Hawking's original analysis,[22] this process takes place because part of the information is swallowed by the black-hole's singularity. Hawking argued that this process would continue until the eventual disappearance of the hole in a "pop" (an explosion in which the hole's final mass of only about 10^{-5}g is converted into energy), and that the information, swallowed by the hole's space-time singularity, must finally be considered at be lost to the universe.

In more recent years, many physicists (and eventually even Hawking himself) have tried to argue that the information is somehow restored through such effects as subtle correlations in the Hawking radiation. However it has long been my opinion, and has also been strongly argued by others[23], that this is not correct and that the information is indeed lost. To put the matter more correctly, there is a loss of phase-space that is available to the universe, during the swallowing of information by the black hole, and the relevant degrees of information become completely inaccessible to the universe once the black hole has finally evaporated away in Hawking's "pop". This leads to a "thinning down" of the universe's total phase space, so that as measured on the scale of the universe as a whole, when a black hole disappears the same coarse-graining region as before, as far as the rest of the universe is concerned, will count as having a smaller volume (and indeed a smaller dimension) than it had before. In CCC, this thinning down of phase space volume eventually completely compensates for the increase in entropy that had occurred throughout the aeon's history, so that the hugely reduced degrees of freedom that are available in the aeon's final exponentially expanding state can match completely with those that are available at the big bang of the succeeding aeon!

Clearly CCC raises many issues here which need detailed checking for their internal consistency. And there are many others which will have to stand up to the rigours of observation and future experiment. It will be of great interest to see how all these matters develop in the future.

Acknowledgements

The author is grateful to many colleagues and acquaintances for valuable criticisms and contributions to the ideas described here; he is also grateful to NSF for support under grant PHY00-90091.

References

Bjorken, J. and Drell, S. (1965) *Relativistic Quantum Fields*, McGraw-Hill. ISBN 0-07-005494-0

Braunstein, S.L. and Pati, A.K. (2007) Quantum information cannot be completely hidden in correlations: implications for the black-hole information paradox, *Phys. Rev. Lett.* **98**, 080502.

Friedrich, H. (1986) On the existence of n-geodesically complete or future complete solutions of Einstein's field equations with smooth asymptotic structure. *Commun. Math. Phys., 107*, 587-609.

Guth, A.H. (1981) Inflationary universe: A possible solution to the horizon and flatness problems. *Phys. Rev.* **D23**, 347-56.

Hawking, S.W. (1975) Particle creation by black holes *Comm. Math. Phys.* **43**, 199-220.

Hawking, S.W. (1976a) Black holes and thermodynamics *Phys. Rev., **D13(2)*** 191-

Hawking, S.W. (1976b) Breakdown of predictability in gravitational collapse *Phys. Rev., **D14*** 2460-

Kronheimer E.H. and Penrose, R (1967) On the structure of causal spaces. *Proc. Camb. Phil Soc.)* **63**, 481-501.

Misner, C.W. (1968) The isotropy of the universe. *Astrophys. J., **151***, 431-57.

Penrose, R. (1965) Zero rest-mass fields including gravitation: asymptotic behaviour, *Proc. Roy. Soc. London*, **A284**, 159-203.

Penrose, R. (1979) Singularities and Time-Asymmetry, in *General Relativity: an Einstein Centenary*, eds. S.W. Hawking and W. Israel (Cambridge University Press).

Penrose, R. (1990) Difficulties with inflationary cosmology, in *Proceedings of the 14th Texas Symposium on Relativistic Astrophysics* (ed. E. Fenves, *N.Y. Acad. Sci.*, New York).

Penrose, R. (2004) *The Road to Reality: A complete guide to the Laws of the Universe* (Jonathan Cape, London).

Penrose, R. (2005) http://www.newton.cam.ac.uk/web seminars/pg+ws/2005/gmr/ gmrw04/1107/penrose/.

Perlmutter, S. *et al* (1998) Cosmology from type Ia supernovae *Bull. Am. Astron.* **29**. [astro-ph/9812473]

Rindler, W. (2001) *Relativity: Special, General, and Cosmological* (Oxford University Press, Oxford).

Rugh, S.E. and Zinkernagel, H. (2007) Cosmology and the meaning of time (Symposium, 'The Socrates Spirit' Section for Philosophy and the Foundations of Physics, Hellabaekgade 27, Copenhagen N, Denmark).

Steinhardt, P.J. and Turok N. (2001) A Cyclic Model of the Universe http://xxx.lanl.gov/PS_cache/hep-th/pdf/0111/ 0111030.pdf

Steinhardt, P.J. and Turok, N. (2007) *Endless Universe: Beyond the Big Bang* Doubleday, New York.

Tod, K.P. (2003) *Class. Quantum Grav.* **20** 521-534.

Veneziano, G. (1998) A simple/short introduction to pre-big-bang physics/cosmology, hep-th/9802057 v2 in the LANL Archive

Endnotes

[1] Sections 2, 3, 4, and 5 of this article are taken almost *verbatim* from Penrose (2005), with the kind permission of the publishers.

[2] The basic idea of CCC first came to me in August 2005. Comparison should be made with the proposals of Veneziano (1998) and of Steinhardt and Turok (2001, 2007). However, these other schemes differ substantially from CCC and depended upon ideas taken from string theory.

[3] See Perlmutter *et al.*(1998).

[4] See, for example, Kronheimer and Penrose (1967).

[5] See Guth (1981).

[6] I am using the capitalized form "Big Bang" for the specific event that initiated our aeon; "big bang" is used here more generally for the corresponding event for any aeon.

[7] See Misner (1968), for example.

[8] Penrose (1990, 2004 Chapter 27).

[9] Phase space is a mathematical space P, normally of an enormous number of dimensions, in terms of which a single point of P encodes the entire configuration of a physical system together with the instantaneous motions of all its parts (given in terms of their individual positions and momenta); see, for example, Penrose (2004), Chapter 20.

[10] In cosmology, a "co-moving" region refers to a part of space-time that follows the history of the motions of the (idealized) galaxies; see, for example, Rindler (2001).

[11] See Hawking (1976a). Here k is Boltzmann's constant, G is Newton's gravitational constant, h is Planck's constant, and c is the speed of light.

[12] See, for example, Penrose (2004) Chapter 27.

[13] Penrose (1979).

[14]Tod (2003).

[15]Penrose (1965).

[16]See also Rugh and Zinkernagel (2007).

[17]See Penrose (1965).

[18]For example, Perlmutter *et al.*(1998).

[19]This is a reasonable conclusion from the work of Friedrich (1986).

[20]It follows from quantum field theoretic considerations of pair annihilation that there cannot be any massless charged particles around today. See Bjorken and Drell (1965).

[21]See Penrose (1965).

[22]Hawking (1975, 1976a,b).

[23]See Braunstein and Pati (2007).

CHAPTER EIGHT

Deconstructing Deathism:
Personal Immortality As A Desirable Outcome

R. Michael Perry[1]

Grapes, Sour and Sweet

In Aesop's ancient fable, the fox seeks the juicy grapes to quench his thirst on a hot, sunny day. Finding them out of reach, however, he concludes "they must be sour."

The thirst for longer life and better health, which would hopefully extend to a happy existence of indefinite duration, is basic to human nature. Just about everyone has been tempted by these appealing "grapes," notwithstanding that a substantial extension of maximum human life-span, healthy or not, is quite out of reach at present, and always has been. Mortality is a basic feature of earthly life. Yet humans, who seem to be the first life forms on the planet to understand this, are not happy with it. Yes, death is "natural," but our instincts tell us it's still not "okay."

The roots of our irrepressible anti-death wish stretch well into prehistoric times, as is suggested, for example, by the burial of artifacts such as hunting implements with the dead. In more recent though still ancient times, the feeling flowered into major religions that promised the sought-for immortality and a happy future existence. Many of these belief systems are still with us and their adherents total perhaps about half the humans alive today. We see then how the wish for existence beyond the biological limits has survived the intractable difficulties that its practical realization has offered. In recent years, moreover, hopes for death-transcendence have taken on new life through scientific advances that offer possibilities of addressing the problem directly. The mechanisms of aging are being unraveled and eventual, full control of the aging process and known diseases is anticipated by some forward-looking people, along with other life enhancements not previously known. People can meanwhile arrange for cryopreservation in the event of death, in hopes that resuscitation technology will eventually be developed, along with the means to reverse or cure any affliction they may have suffered, including aging itself.

Not everyone, of course, can be counted among the immortality-seekers or supporters, even when the new scientific perspective is taken into account. Among those who freely reject the "grapes" of life extension are a predictable fraction who would find them sour as well. These critics defend a counterproposal of deathism, namely, that not only is one's eventual demise inevitable and final (the grapes are out of reach) but that this should be seen in a positive light (but sour too, so not to worry). An essay in this vein which I have especially focused on here is *The Immortal's Dilemma: Deconstructing Eternal Life* by George Hart.[2] It offers the opinion that "life can have meaning only if it must end" and argues the case against the prospects for immortality on logical grounds. Such criticism is useful, for it points up difficulties that must be solved if immortality is ever to be realized. On the other hand, the *possibility* that immortality can be realized, and realized as a desirable and rewarding endeavor for an individual life (so the grapes are reachable and sweet and juicy after all) is not refuted by such arguments, as I shall maintain here. (I also confess to being among those whose hopes rest on these grapes being in some way reachable, with emphasis, in my case, on scientific approaches to the problem.) In addition to Hart's own critique I will also consider more briefly some other deathist arguments that have made their appearance over the centuries. But first some comments are in order about what I think immortality should encompass.

Here I am largely in agreement with Hart himself who (along with many others who have commented on the issues) is not only a materialist and a rationalist, but is also sensitized to certain difficulties of an informational nature that, I think, especially must be addressed. (Technically I classify myself as a physicalist rather than a materialist, holding that reality is explainable by physics, whatever one thinks about the nature of matter and material objects versus other possible ontological substrates such as information.) Thus I discount any idea of immortality "outside of time" or any supernatural or mystical process or entity taking part. A person, to exist at all, must always remain part of physical reality as revealed and understood scientifically. I also discount any idea of immortality, whether scientific or not, based on attaining a "final" mental state or a limited repertoire of states and remaining in that condition without significant change. That would amount to what is called an Eternal Return, in which one has only a finite number of subjective experiences, even if repeated endlessly—not true immortality in my view (or Hart's, once again). An immortal life must avoid this problem of *stagnation*, instead becoming an endless process of *personal growth* which, among other things, would allow for continual recall of a growing body of past experiences. Endless personal growth would mean our immortal is continually changing—though not arbitrarily. Actually, this will cause certain arguments against immortality that easily come to mind to lose force, as we shall see, though also raising an additional, challenging problem.

It is worth remarking here that a suitable habitat for endless growth would have to exist, an expanding or already-infinite domain. Ultimately it would seem to resolve into whether information encoding memories, dispositions, and a general record of the past characteristics of the individual could be suitably recorded and organized on an ever-expanding scale. It is not known at present whether our own universe, though it appears to be expanding, could support such a process, but the possibility is not ruled out. To reasonably accommodate one immortal being, such a growth process should also extend to an entire, large population of developing immortals, so that each individual is progressing in more-or-less similar fashion. (This would also allow the addition of new, developing individuals from time to time in unending succession, though the rate of addition, as well as the growth itself, would have to be managed to be consistent with available resources. I should also add that the growth process of each person could survive temporary reversals including some losses or corruption of information, so long as overall trends were suitably robust. Basically, a subset of the information taken in by the individual should accumulate without limit and never be permanently lost or altered.) We shall return to this subject briefly later, in connection with the idea of multiple universes, which, if accepted, will be seen to further strengthen the prospects for some form of immortal habitat.

The developing immortal, then, would acquire experiences which would from then on be available for recall. Such recall would have to happen repeatedly, otherwise a given experience would drop out of consciousness forever at some point and not be part of that individual. A growing body of experiences would have to be recalled or reviewed infinitely often over infinite time to avoid stagnation. This, however, will be seen to raise a further difficulty, as Hart also notes, a problem of *dilution*. An experience or set of experiences might be very seldom recalled even if the recall is infinitely often, in view of the growing body of other material demanding attention. In this way substantial portions of one's past, or ultimately all portions, may, for practical purposes be lost from consciousness and not part of the "self." But I will argue that this problem too is manageable or at least cannot be shown not to be. Thus one could either cultivate a tolerance for an increasingly infrequent recall of a given past experience, or actually eliminate the problem by a suitable scheduling of the time spent reviewing personal archives.

We shall now examine Hart's main arguments in more detail.

Stagnation and the Death Wish

An immortal being must persist for an infinite length of time. Hart argues that, during that interval, such a being must at some point find life unbearable and wish to die, and indeed, by implication this must happen infinitely often. So, even if one always changed one's mind later and again wanted to live, an ordeal of misery and frustration would have occurred, and moreover, must recur, over and over, infinitely often. Why is this? "It is logically possible," he says, "and given our nature as human beings, it is also empirically possible." On this basis he concludes that, "[g]iven an infinite period of time, what remains possible during that period of time is certain to occur." His reasoning is that "[a] possibility that remains open by definition is certain to happen given enough time; otherwise it is meaningless to say that it remains an open possibility if it might never happen even in an infinite period of time."

My answer to this starts with the concession that, since even an immortal being must be subject to the physical laws that govern reality, the wish to die must remain both logically and empirically possible throughout time—here I agree with Hart. Yet the conclusion that such a wish *must* occur (and must recur) is still fallacious, because of the assumption of personal growth which, as we noted, is necessary to avoid stagnation. A person, seen as a developing entity, would not simply be a static construct with fixed probabilities of certain things happening. With a fixed probability an event of given type, assumed independent of other events of the same type, is guaranteed to happen eventually, according to a predictable scheme. For example, suppose a devastating flood has a one percent chance of happening in any one year in a certain locale whose topography is assumed to be fixed. Then the chance of its happening over 100 years is about 64%, and the chance of its happening at least once in 1,000 years is about 99.995%, that is to say, near certainty. (For longer time intervals we come ever closer to perfect certainty.) But by taking proper precautions it would be possible to change the relevant probabilities so that an undesirable occurrence such as this becomes increasingly unlikely. People could, for instance, shore up a system of levees (slightly changing the "fixed" topography) to make a bad flood less likely, and might do so repeatedly or make other changes to further reduce the likelihood.

In the case of wanting to die, one would naturally be interested in reducing the likelihood of such a state of despondency (or reckless curiosity?). Furthermore, the sort of personal growth I envision, which would encompass the whole of a large population, should result in ever-increasing, widespread levels of intelligence and capability to deal with problems of all sorts. This is not to say that problems will not occur and persist, and in fact, some problems could become more acute with the increasing levels of sophistication, much as humans today may be said to have more in the way of psychological problems than an earthworm. But certainly the prospect of dealing successfully with the problems cannot be ruled out. So, for example, our immortals could get happier and happier, or more and more firmly resolved to stay the course of living, or both. The likelihood, after a certain point, of a suicidal impulse *ever* occurring could then be vanishingly small, even though it would never drop strictly to zero.

As an illustration, we may imagine that at some future time the probability of a serious suicidal spell has been reduced to one percent per annum, and that it undergoes a further, exponential decay over time, with growing enlightenment and overall good feeling. With a half-life of 100 years, so that the probability reduces by half every century (though again, never going all the way to zero), the probability of there ever being such an episode is not 100 but only 77 percent. A half-life of 50 years will bring the probability down to 52 percent, and one of 30 years will cut it to 35 percent. Going back again to the case of the 100-year half life, the chance of at least one bad episode happening in 1,000 years is very nearly the same as its ever happening at all, or about 77 percent, but the chance of its happening after this first 1,000 years is minuscule, only about a thousandth of its ever happening at all, or .08 percent. With a 50-year half-life, the chance of a bad episode in 500 years is similarly very close to the 52 percent figure for all future time, but the chance of its happening after the 500 years is again a thousandth, in this case, .05 percent. And so on. We see then how a favorable outcome—no bad episodes at all over infinite time—becomes a near certainty with the passage of time, even though there is always some tiny chance of the contrary. (I will add that here we have assumed an exponential decay of probabilities, which makes calculations easy, but such a specific falloff is not essential; many other falloff curves will do as well.)

The same sort of argument could be applied against other "inevitable" consequences such as simple physical destruction. Developing individuals will naturally occupy larger and larger volumes of the universe, or at any rate, a larger volume in some cyberspace storing information (extending ultimately to larger spatial volumes). They thus should be able to make themselves progressively immune to such destruction, through storage of backup information and the like, even though a minuscule and diminishing probability of such destruction will always remain.

Dealing with Dilution

The second major argument Hart raises against the feasibility of immortality invokes what I have called the problem of dilution. Basically, the growing individual must eventually dwarf any previous version of itself, both informationally and, since information requires storage space, physically as well. One consequence is that, in one way or another, an immortal must develop far beyond the present human level. It is easy to see how this could create problems, though as usual we must also ask if these problems must *necessarily* be insoluble.

The main problem would seem to be that of a simple outweighing of earlier information and thus, of the characteristics that defined one's identity at a particular age. The first century of the life of the individual, for example, will be represented by a finite record of, say, N bits. This archival record must occupy an increasingly small part of the total information content of the individual, say it is M bits, as growth occurs and M increases. (The N bits could also be copied repeatedly over time as insurance against loss, but would still at most amount to N bits of nonredundant information.) In time the N bits will be an utterly insignificant portion of the M bits, say a trillionth part or less. It is an easy conclusion that the significance to the individual of the N bits must be correspondingly tiny. In other words, the first century of your life will be as nothing to what you will have developed into—so the early person—including yourself today—will essentially be dead—even though information to reconstruct this version of you still survives. (Doing that, however, would not solve the problem long-term, because dilution would only recur as the new instance of "you" developed and accreted information. Trying to keep "you" alive by periodic recreations, on the other hand, would not work either, because of the problem of stagnation—"you" would just run through a limited repertoire of experiences before dilution once again set in and shut "you" down.)

But wait a minute—must we conclude that dilution would have to be such a problem? Surely not, if we imagine our advanced person has a certain respectful attitude toward the full collection of its past information, and the relationships between the various parts, forming a coherent whole. (This would recount both good and bad times, capture emotional as well as factual content, and be valued for lessons learned through sometimes painful mistakes along with remembered enjoyments.) A librarian does not necessarily think less of the books already on the shelves even when many more titles are acquired. This might hold all the more if the librarian is a scholar who has assembled a well-organized personal library of specially valued books that are consulted and studied from time to time. The scholar may in turn be a historian, and the "books" may include manuscripts and other memorabilia which provide information about historical periods of interest. True, if the library is extensive it may take a while before a given item in the collection is consulted once more, but that would not make it acceptable to discard that item, or necessarily lessen the item's influence on what the scholar is doing. Finally, if we suppose that some or much of that history is personal history, our "scholar" is starting to resemble our hypothetical immortal. In short, we are not justified in assuming that an infrequent perusal of information necessarily negates the importance of that information in whatever manner it is used, including the complex activities that might be involved in expressing and experiencing one's identity.

Today we consult books in a library by physically lifting them off the shelf and opening them up, but that is beginning to change with electronic data bases, which can be scanned much more rapidly by computer. In the future it should be possible for us to scan our own memories much more rapidly and reliably than at present, to lessen the time between scans of particular archival material. At the same time, as we grow our thought processes should also deepen, so that more in the way of processing will be required for many commonplace mental activities. This in turn would offer more opportunity for interleaving the occasional references to times past which will better anchor our sense of who we are by reminding us of where we have come from.

It seems reasonable that past versions of the self would "survive" as we remember the events of times past, that is to say, as we review our episodic memories. This would have importance in our continuing to persist as what could be considered the "same" albeit also a changing, developing person. But in addition to this mnemonic reinforcement I imagine there would be a more general feeling of being a particular individual, an "ambience" derived from but not referring to any specific past experiences. Ambience alone would not be sufficient, I think, to make us who we are; episodic memories would also be necessary, yet it could considerably lessen the need for frequent recall and thus alleviate the problem of dilution.

Another interesting thought is that certain items might consistently be consulted more frequently than others. (Indeed, would this not be expected?) In this way it would actually be possible to bypass the dilution effect and instead allow a fixed fraction of time for perusal of any given item, even as more items were added indefinitely. A simple way of doing this could be first to allow some fixed fraction of the time for day-to-day affairs and other non-archival work ("prime time"), then spend the rest of the time on perusal of personal archives ("archive time"). The exact apportioning of prime versus archive time is not important here, but it will be instructive to consider how the archive time itself might be subdivided. A simple, if overly simplistic, strategy would be to have half this time devoted to the first century's records, half the remainder to the second century, and so on. (Since there would only be a finite number of centuries, there would be some unused archive time at the end, which could be spent as desired. Note, however, that in the limit of infinite total time covering infinitely many centuries, the usage of archive time would approach but not exceed 100%.) In this way, then, there would be a fixed fraction of archive time, 2^{-n}, spent on the nth century's records, regardless of how many centuries beyond the nth were lived or how many records accumulated. True, this way of apportioning time might not be much good beyond a few centuries; only about one trillionth the total time would be spent on the 40th century, for instance, around 1/300 sec per 100 years.

(Possibly a lot could be covered even in this brief interval of about 3 million nanoseconds, however.) But the apportionment scheme could be adjusted.

A more interesting and plausible, if slightly harder-to-describe scheme would be to choose a constant $c > 0$ and allow the fraction $c(1/(n+c-1) - 1/(n+c))$ to the nth-century records. It is easy to show that the time for all centuries will add up to 100% as before, whatever positive value of c we start with. Starting with $c=10$ will get 10% of the total time spent on the first century, with subsequent centuries receiving a diminishing share as before, but the rate of falloff will be much slower, so that the 40th century will still receive 0.4%, or about 5 months per 100 years, that is to say, 240 million nanoseconds per minute. If we suppose that our immortal settles eventually into a routine in which 10% of the time overall is archive time, there would be 24 million nanoseconds available each minute of life for the 40th century's memories alone, if desired, with many other centuries getting more or less comparable or greater amounts of attention, and, as usual, none omitted entirely. This, I think, makes at least a plausible case that a reasonable sense of one's personal identity could be sustained indefinitely.

In the above examples the greatest proportion of archive time falls to the earlier records, which might be fitting since these should be the most important as formative years for the prospective immortal, thus the most important for identity maintenance. (Memory recall would also naturally occur during prime time; the emphasis here could be on recent events, to maintain a balance overall.)

In summary, we have considered ways that the problem of dilution might be successfully managed. Relatively infrequent perusal of memories might still suffice to maintain the necessary continuity with past versions of the self, or proper scheduling could stabilize the frequency of recall and bypass the dilution effect, or both. We see in any case that the problem is not what it may seem at first sight.

We have no guarantee, of course, that it would not get out of bounds, but certainly some grounds for hope.

More could be said, but the difficulties are formidable, trying as we are to anticipate the possible future before it happens, and how we will deal with our problem of memory superabundance when many new options should have opened up. In that hopefully happy time a "science of personal continuation" should have taken shape to properly deal with the matter. Nay-sayers like Hart try to discount any such prospects once and for all, based on today's perspectives with their inevitable limitations. We must look to future enlightenment to overturn such summary judgments. I will have a bit more to say on this issue, in the process addressing some other notable pro-death thinking. But first it will be worthwhile to consider a few additional points raised in Hart's essay. These again I think offer no fundamental, demonstrated difficulties to the idea of immortality.

Earlier we noted Hart's bringing up the problem that the would-be immortal may at times undergo a change of feeling and wish for death. While I think we have disposed of his claim that the death wish, to remain an open possibility, would have to actually occur and recur at a serious level, it is also significant that he would allow the option of suicide, supposing such a wish did occur. And here I agree with him, if reluctantly, since a person should have that right. As an aside he seems to think of choosing to be "mortal" as an alternative different from suicide, though he does not explain how. To kill oneself with a slow-acting poison or microbe would still be suicide; would that not hold even if the process took decades and is now "natural," as in the aging process? Choosing to age and die as we do today at a time when aging could be reasonably controlled and prevented strikes me as a suicidal choice. But, in fairness to Hart, the delay could have significance inasmuch as the subject could undergo a change of views meanwhile, and opt for a reversal or cure. More generally, though, the rather morbid dwelling on a putative, recurring death-wish suggests that Hart may not be so happy with his own life but may instead in some degree be yearning for an "honorable" way out. Such an outlook is all too common among people, intelligent thinkers included. All such people should take seriously the prospect of becoming joyful geniuses—or of enhancing their already-existing genius and joyfulness—which future advances should make increasingly feasible.

True, many such people might object that doing this would make them so different it would no longer be *them*, they would be dead for all intents and purposes and the new person would be someone else. But I seriously doubt this would have to be so, and wish I could persuade these nay-thinkers to give more thought to the matter. A change of mind and heart need not add up to a change of person, with the old dead and gone, but can also be seen as a fulfillment of the old, which is thereby helped to become better than before, continuing in its progress find ever-deeper meaning and greater joy.

Morbidity and Its Remedies

The impression of morbidity in Hart's thinking is reinforced by his opinions on very long life. "In theory you can imagine without contradiction what it would be like to be alive for a trillion or even a trillion trillion years from now. This thought experiment creates its own horror, one that is mind-numbing and nauseating." Personally, I find the thought experiment not nauseating but exhilarating! What incomparable wonders one might explore in such long periods, what fascinating problems one might solve! What endearing relationships one might have with others of sympathetic but still differing minds, what great good one might do, with reciprocal rewards for the well-enlightened! Hart offers the thought that life ought to be like a book, which has a beginning, middle sections, and an end. In this way one's life is "properly framed," says he, and only in this way can it have meaning. A big problem I see with his analogy is that, while you can appreciate the "framing" and thus the meaning of a book by reading it through to the end, to do it right requires some thoughtful deliberation after you have finished the book. This is not an option you can exercise with your own life, if it too must come to a final stopping-point.

The dreary thought that one's life needs a "conclusion" seems wrong and misguided to those of us who would like it to continue without end. (A life rightly lived is never rightly, permanently ended, we say in earnest rebuttal.) Yet it does beg the question of what meaningful activity would demand and occupy an infinite future, one in which we can and must progress indefinitely, yet continue always to respect and, in some appropriate measure, identify with our much humbler beginnings. How would an infinite existence be made worthwhile and necessary? Certainly it sounds like a tall order, but is it such an impossibility, assuming of course that the necessary technological advances will occur to at least permit escape from the biological limits that now confine us?

Indeed, from one point of view the issue seems transparently simple. Life ought to be worth living. If life is worth living, it should not come to an end, therefore one ought to be immortal. This, of course, overlooks the details of what one might be doing with one's life as well as such other features as what sort of society would emerge if individuals were immortal. These matters are impossible to second-guess in detail, but some things can be said with reasonable confidence.

Whatever the details of a life may be, they should be such as to produce meaning and fulfillment—including, most importantly, a reason to continue, to find something always new, interesting, exciting, something from which one can learn. This applies to our limited existence today; it should apply all the more in a hoped-for immortal future. Life should be habit-forming! With the prospects for future betterment, I think it will be, both because there should be so much of interest to experience and know about, and because our means to deal with the problems of lack of interest and other negatives will itself be much greater and more refined.

Another aspect of life being worth living is that it should be worth remembering, as we have already noted, this in particular being necessary to retain a sense of continuity with one's past to reasonably sustain one's personal identity. Pleasure alone thus is not enough. The nature of one's experiences should be such that thinking of them later causes enjoyment too—a requirement that, I think, should not prove too difficult in the sort of future that seems possible, even though people today often do not seem to value the remembered past.

Finally, what is worth remembering is also worth sharing. Life should be something shared with others so that all in the end will mutually benefit. Of course it must be the "right" others, which will follow if individuals are well-disposed and develop in reasonable ways, a possibility that I conjecture is open to all individuals, even those who presently seem ill-disposed.

So we see that commonsense notions that apply to life today, even with its present limitations, lead to the conclusion that immortal life, properly conducted, would be good and desirable. This is also bolstered by considering the opposite viewpoint. Could we learn to make peace with death? Could we see in it something other than final ruin and frustration? Could we find meaning in spite of (or because of) the thought of an eventual, permanent conclusion, a restitution once and for all of all our striving and cares? I think all attempts to do so must ring hollow. Knowledge of one's mortality and its apparent inevitability is not an easy burden for the rational mind to carry. I doubt if belief in one's impermanence can inspire much real satisfaction, except perhaps for those who view life, fundamentally, as a burden that ought to end. As one such thinker, Hart is hardly alone; a few of the others will now be worth examining, starting with the ancients.

Other Deathist Thinking

The Stoics, prominent in the early centuries C.E., insisted that fear of death, rather than death itself was the real evil, so that "man must learn to submit himself to the course of nature."[3] Now, of course, we know that our nature is substantially malleable through our own efforts. The sort of meek submission advocated in earlier times is becoming untenable, and increasingly will be so.

The related, roughly contemporary Epicurean doctrine held that stagnation would invalidate a limitless survival. "[T]here are only a limited number of gratifications, and, once these have been experienced, it is futile to live longer."[4] To me, this conclusion seems especially specious, even if we limit consideration to a purely intellectual discipline such as mathematics. There are infinitely mathematical truths to explore, each a separate and unique "gratification" to the rightly disposed, with no simple way to characterize them all—Gödel's famous undecidability results establish this property about as solidly as one could ask. Again, too, our nature is malleable thus allowing for increases in the "number of gratifications" along with other enhancements, to track the reality that obligingly refuses to be trivial.

Buddhism, also very ancient (and still quite active today), considers the "wish for continued existence" a form of "defilement."[5] This, then, is a moral objection to immortalism, one with which we may respectfully disagree. (Though perhaps we should examine carefully the intended meaning of "existence' here—I assume a straightforward interpretation implying survival as considered here, though this is not the only possible interpretation.) Buddhism strongly advocates enlightenment; more enlightenment should be possible the longer one lives. In time, I conjecture, such enlightenment will lead to a recognition of the individual person as a coherent concept and something whose continued existence is to be valued and sought.

Turning to more recent times, Bertrand Russell, a leading twentieth-century British philosopher, was firmly convinced of the inevitability of death, based on cosmological considerations. If nothing else, he thought, life must eventually and uniformly come to an end in the Heat Death or "running down" of the universe. Not just individuals were doomed but species, civilizations, and in short, the whole enterprise that we know as life, whether earthly or elsewhere in our cosmos, if it should exist there. Russell was not happy with this state of affairs but thought it must be accepted, arguing that "...only on the firm foundation of unyielding despair, can the soul's habitation...be safely built."[6] His solution was to downplay the issue. The thought that "life will die out...is not such as to render life miserable. It merely makes you turn your attention to other things."[7] But this too rings hollow in the minds of many of us. In particular, it invites the question of whether painless, immediate suicide would not be a better alternative than prolonged and distracting efforts at "other things." Russell does deserve credit for attempting to assess reality as it is, and make the most of what to him inspired "unyielding despair."

It is worth remarking that Russell's conclusions about eventual Heat Death with its apparent stifling of all life in the universe have never been ruled out but are not by any means firmly established. The recent astronomical finding of an apparently accelerating universal expansion has raised new, unanswered questions about the ultimate fate of the universe and any firm conclusions are premature. On the other hand, if we suppose the universe is destined to end or otherwise prove fatal to life within it, we can still ask if this is the absolute terminus. Barring the supernatural, many would say yes. However, suppose we accept the idea of pattern survival—that "you" could survive as a duplicate of yourself, possibly located in another universe entirely (and no one has ruled this out). Then clearly the options for survival are broadened so that even a hostile cosmology may not be able to end your existence. Along with this prospect would come that of resurrecting the dead in replica form, something we might conjecture must be inevitable if reality as a whole has no finite limits. Life, not death, could be the ultimate outcome for any individual, who must then make the most of it rather than seeking solace in a cares-erasing oblivion.[8]

John Hick, a prominent contemporary theological philosopher, has also aired misgivings on the issue of eternal survival. His difficulty is a variant of the problem of dilution. There must be a limit, he says, to how much we can identify with earlier states in which we were very different. In addition to logistical difficulties of the sort we addressed earlier, Hick considers the diary he composed as a fifteen-year-old (emphasis original): "...I know that it is *my* diary, and with its aid I remember some of the events recorded in it; but nevertheless I look back upon that fifteen-year-old as someone whose career I follow with interest and sympathy but whom I do not *feel* to be myself."[9] This sort of dissociation is, I think, very common and perhaps a majority viewpoint among people today, though not universal. (I for one feel able to identify with my earlier person-stages, even going back to early childhood, despite the many changes that have occurred.) It is noteworthy that Hick says he does not *feel* he can identify with his earlier self.

It is not likely that any of the arguments offered here would soon change such a viewpoint. If we must continually change so that, in time, our earlier experiences were of someone very different this might indeed prove a fatal impediment, but I do not think it must or will be so. The arguments we have already considered offer a starting point for a more hopeful outlook, but we can go a bit further informally, something I find inspirational. Let us consider, then, what sort of beings we might be expected to develop into over a long stretch of time, in which today's physical limitations would not apply.

Likeable, Joyful Immortals All

Clearly there are many possibilities, but I conjecture that personality types capable of and desiring infinite survival will not be so varied or inscrutable as to baffle our understanding today. Instead they should be profoundly benevolent, desirous of benefiting others as well as themselves, and respectful of sentient creatures in general. They will acknowledge that enlightened self-interest requires a stance with a strong element of what we would call altruism. They will see that, to obtain maximum personal benefit, interests of the community at large must be considered, and the highest happiness, to be enjoyed by some, must be shared by all. They will be intensely moral, but also joyful in the exercise and contemplation of their profound moral virtues—for a substantial element of joy will be essential in finding life worth living, even as it is today. These joyful, good-hearted beings, then, will be the types to endure, and will refine their good natures as time progresses, so as to increasingly approximate some of our ideals of angelic or godlike personalities, as endless wonders unfold to their growing understanding.

Beings of good will who are seeking what is right and best and to develop in wonderful and rewarding ways over unlimited time, always with love, respect, and consideration for others, should not find it hard to feel a kinship with past versions of themselves which also had these attributes. Love must conquer all. The conjectured disinterest with one's more distant past, then, will be swallowed up in the universal affection and regard for persons in general, past as well as present, which must logically extend to versions of oneself along with others. If we are good enough, then, our everlasting survival, as separate though interacting and considerate selves, becomes morally mandatory and recognizable as such by the advanced beings we shall become. So it is this high calling we must aspire to, and it may well be necessary to our survival. And, I submit, being virtuous and considerate will also make us more accepting of our earlier selves, even if they were less enlightened and rather "different," or even, in more extreme cases, profoundly evil and horribly misguided. The bad in our earlier selves can be acknowledged when we are confident it is cured.

In the future there should be wonders aplenty for the searcher and many paths to pursue in a vast architecture of possibilities. So each of us should be able develop in interesting and unique ways, with joy accompanying our efforts, including those occasions when we reflect on where we have been before and how far we have come, something that should both comfort and inspire. Joy will thus help us maintain a reasonable sense of our identity as time goes by. If this course of development can be pursued, the rich diversity of individuals will, I submit, produce greater benefits overall than if all were subsumed in a vast collective enterprise, with individuality devalued or obliterated. As a possible precedent, we may consider how collective enterprises in our own history, and particularly totalitarian governments with centrally planned economies, have been unable to compete with more decentralized, democratic systems. A happy medium must be found, however, between extremes of collectivism and egoism. The separate, developing, considerate, immortal ego, then, should have more to offer than some form of "nonself" or a fused consciousness, both to itself individually and the community at large.

In our advancement, of course, we should make use of whatever discoveries and technologies may be applicable. Inevitably this will involve risk but "nothing ventured, nothing gained." In fact I think our deepening understanding will make adaptations possible that would otherwise be out of the question. The elimination of aging and biological death should be accompanied by increased understanding of the psychological difficulties connected with immortalization, with a proliferation of possible remedies.

People should have numerous means to deal with various "illnesses" they may have inherited from the mortal past, and even death itself, along with the difficulties they encounter in the course of a hopefully unbounded future.

Bibliography

Gruman, Gerald. "A History of Ideas about the Prolongation of Life." *Transactions of the American Philosophical Society* 56, no. 9 (December 1966).

Hart, George. "The Immortal's Dilemma: Deconstructing Eternal Life," Internet Infidels, Inc., 1973: http://www.secweb.org/asset.asp?AssetID=333 .

Hick, John. *Death and Eternal Life*, Louisville, Ky.: Westminster/John Knox Press, 1994.

Perry, R. Michael. *Forever for All: Moral Philosophy, Cryonics, and the Scientific Prospects for Immortality.* Parkland, Fla.: Universal Publishers, 2000.

Saddhatissa, Hammalawa. *Life of the Buddha.* New York: Harper and Row, 1976.

Tipler, Frank J. *The Physics of Immortality: Modern Cosmology, God, and the Resurrection of the Dead.* New York: Doubleday, 1994.

Endnotes

1. An earlier version of this chapter appeared in *Physical Immortality* **2**(4) 11-16 (2004), and (subject to updating), at http://www.universalimmortalism.org/deconstructing.htm .

2. George Hart, "The Immortal's Dilemma."

3. Gerald Gruman, "A History of Ideas about the Prolongation of Life," 15.

4. Ibid., 14.

5. Hammalawa Saddhatissa, *Life of the Buddha*, 31.

6. Bertrand Russell, *Why I Am Not a Christian*, 107, as quoted in Frank J. Tipler, *The Physics of Immortality*, 69.

7. Ibid., 11, as quoted in Tipler, *Physics of Immortality*, 70.

8. See R. Michael Perry, *Forever for All*.

9. John Hick, *Death and Eternal Life*, 410, as quoted in Perry, *Forever for All*, 470.

CHAPTER NINE

What Mary Knows:
Actual Mentality, Possible Paradigms, Imperative Tasks

Charles Tandy

Introduction

In part one (of two parts) I show that any purely physical-scientific account of reality must be deficient. Instead, I present a general-ontological paradigm. There is reason to believe that this paradigm is acceptable to most persons and philosophers. I believe this general-ontological framework should prove fruitful when discussing or resolving philosophic controversies; indeed, I show that the paradigm readily resolves the controversy "Why is there something rather than nothing?"

In part two, now informed by the previously established general ontology, I explore the controversy or issue of immortality; the focus is on personal immortality. The analysis leads me to make the following claim: Apparently the physical-scientific resurrection of all dead persons is our ethically-required common-task. Suspended-animation, superfast-rocketry, and seg-communities (i.e. O'Neill communities) are identified as important first steps toward the immortality imperative.

PART ONE

Frank Jackson (1982, 1986) gave us a thought experiment now philosophic classic. He has us imagine that Mary the super-scientist was born and raised in a black-white room. We can imagine she was educated with the aid of a black-white library of books and a black-white television-computer; we can imagine that her visitors were dressed in black-white armor; etc.

Such a genius was Mary that she gained all physical-scientific knowledge, including complete knowledge of color vision. Jackson entitled his 1986 article "What Mary Didn't Know" to suggest that when Mary finally steps out of the black-white room for the first time (or is presented with a color TV) – she will utter or think "Wow!" despite her awesome scientific knowledge. For the first time she will know the **experience** of color.

Other similar thought experiments may be constructed to make the same point. Indeed, Jackson (1982) also mentions Fred. Fred sees an extra color unknown to normal humans. (A normal-sighted human person might nevertheless comprehend the science involved in Fred's unusual ability or the special sensory abilities of non-human animals, terrestrial or extraterrestrial.) In 1974, Thomas Nagel had asked "What Is It Like to Be a Bat?" Even if we had all physical-scientific knowledge, we still would not know what it is like to perceive something with the bat's sensory system.

Herbert Feigl (1958) had reminded us yet again of "the alleged advantages of knowledge by acquaintance over knowledge by description. We may ask, for example, what does the seeing man know that the congenitally blind man could not know." [1] P. F. Strawson (1985) discussed the mental and the physical by pointing out that human history can be recounted in two different ways. [2] A physical history might focus on the changing physical position of human bodies or their atoms. But a personal history might explain human action in terms of mentality (e.g. beliefs, desires, or perceptions). Strawson does not see any conflict between the two accounts. Indeed, in 1958 Strawson had written: "What I mean by the concept of a person is the concept of a type of entity such that both predicates ascribing states of consciousness and predicates ascribing corporeal characteristics...are equally applicable to a single individual of that single type." [3]

Mary the super-scientist is a human person; she has a sense of physical reality and of mental reality. In addition, she has a kind of sense of the totality of reality even though she has never seen or experienced the whole or entirety of everything. I will use this account (my account) of what Mary knows to develop one possible ontological paradigm or general ontological cosmology. Both before and after Mary's experience of color, we can say she would make the following distinctions: [4]

1. Mental-Reality
2. Physical-Reality
3. All-Reality

We have also specified that Mary knew all physical-scientific knowledge both before and after her experience of color. But Descartes had attempted to begin developing his ontological paradigm by assuming no such alleged knowledge. We can say with Descartes: I think (in the sense that I am aware that I am thinking), therefore I am (mentality). [5] Beyond that, as a practical matter it seems almost inevitable that any human person (e.g. Strawson) would posit some paradigm or other that included an external reality of physical entities: "objects" (physical nonmental entities) and "persons" (physical mental entities). (My own "actual mentality" has the ability to posit or believe or imagine "possible paradigms" such as this one, and to feel an imperative to act one way rather than another.)

Thomas Kuhn's (1962) historical analysis of the ongoing development of scientific knowledge called our attention to the fact that sometimes we simply add new knowledge to an existing paradigm, and at other times we invent new paradigms. Kurt Gödel had previously established that newer and newer systems of mathematical thought encompassing greater and greater insight are never-ending. [6] I conclude that in the finite temporal world there can be no such thing as absolute and complete knowledge of all of reality. Moreover, I don't see how we could ever know in advance (for all time) that our latest paradigm (no matter how long-lasting and super-sophisticated) would never be superseded.

If Mary had been living for several centuries instead of several decades, she would know that our scientific knowledge changes, sometimes in a revolutionary (new paradigm) way. For example, the Ptolemaic cosmology seemed to work well for a while, then Newtonian cosmology seemed to work better. Today we have Einsteinian cosmology but we now believe it could someday be superseded.

In general, human persons have rather stubbornly over many millennia held on to their realities of mentality, physicality, and allness. We find great variations in the details of their paradigms, however. For example, some have said that physicality is a reality of sorts – but is ultimately an illusion in the allness of things. Below I will assume the reality of the three realms without claiming that ultimately reality is an illusion!

The mental-reality of professional mathematicians seems to tell them that mathematicians are discoverers rather than inventors. $1+1=2$. If there were no mathematicians, no persons, indeed no life at all – it is nevertheless the case that $1+1=2$. This is a part of all-reality. We human persons stubbornly maintain that $1+1=2$ despite it being falsified often. There are many facts against the hypothesis that $1+1=2$. One raindrop added to another raindrop results in (not two but) one raindrop. One unit of a particular liquid may be added to one unit of another liquid to give us something distinctly less or distinctly more than two units of liquid. (I am also tempted to mention the empirically tested and confirmed speed-of-light related "twins paradox" of Einsteinian theory.) To put it another way, in the words of Charles Hartshorne: "A statement [e.g. 1+1=2] thus unfalsifiable absolutely is...incapable of being either true or false – unless it is true by necessity. Since it cannot in any significant sense be false, it also cannot merely happen to be true, but can only be necessary – or else nonsensical." [7] To put it yet another way still, 1+1=2 is not falsifiable and is not a scientific hypothesis; rather, 1+1=2 is a necessary part of all-reality.

Accordingly (according to the ontological paradigm just proposed) it is impossible for all-reality not to exist. The 19th century Russian, Nikolai Fedorovich Fedorov, criticized professional philosophers thusly: [8] "How unnatural it is to ask, 'Why does that which exists, exist?' and yet how completely natural it is to ask, 'Why do the living die?'" "Our attitude toward history should not be 'objective', i.e., nonparticipating, nor 'subjective', i.e., inwardly sympathetic, but 'projective', i.e., making knowledge 'a project for a better world'". "In man nature herself has become aware of the evil of death, aware of its own imperfection." Fedorov's can-do attitude motivates us to look more closely at the issue of death and immortality in order to gain and use knowledge projectively for world betterment. Our general mentality-physicality-allness paradigm may assist us or Mary or me to projectively look more closely at the issue of mortality and immortality.

PART TWO

We may identify several uses of the term **immortality**; here are some examples or possible paradigms: [9]

1. **Einsteinian Immortality.** The world is a time-space whole in which the past (including every human person) will always exist.
2. **Spiritual Immortality.** When my body dies I will nevertheless continue to live; my spirit (as a multi-staged life or career) will continue to have experiences.
3. **Cosmic Immortality.** We came from the eternal cosmos and upon death will return to the cosmic mind.
4. **Physical Immortality.** The pattern of components that constitute my body (e.g. my brain) may be disrupted to the extent that the pattern no longer physically exists; restoration of the pattern to functional physical existence would resurrect me.

We can add additional example paradigms. One may talk of memorial immortality via those who remember you after your death. One may talk of biogenetic immortality via transmission of one's genes to offspring. Reincarnation immortality comes in many varieties but typically tends to be a variation on spiritual immortality or cosmic immortality or both. Christian immortality comes in many varieties but typically tends to be a variation on spiritual immortality or physical immortality or both. It is logically possible for several (all?) of the paradigms cited to be congruent or true concurrently; nevertheless, perhaps you will not find any of the examples personally appealing or motivating once you have removed from the list those you consider unlikely or infeasible. Let me give you my fallible (subject-to-change) take on the possible paradigms (informed by the ontology established in part one above):

Einsteinian Immortality

> Einsteinian Immortality: The world is a time-space whole in which the past (including every human person) will always exist.

I (this writer) first read George Orwell's novel *1984* many years ago (many years before the year 1984). I read the novel because I thought it was science fiction -- but found it to be much more. By the end of the dystopia, even Winston Smith has been thoroughly brainwashed. If his boss holds up his hands saying he has twelve fingers, Winston **actually sees** twelve fingers. Winston **clearly remembers** past events -- but they are events that never **really** took place. This dramatic ending to the novel engrained in me both a sense of the difficulty of uncovering past truths and a belief in the actual existence of the past. (For a detailed philosophic argument that the past will necessarily always exist, see Catterson 2003.)

Once I do X instead of Y, X will **always** be the case. It is impossible for the past to be annihilated; the past necessarily forever continues to exist. It is not just a linguistic convention when we sometimes speak of the past as presently existing. On the other hand, it is not immediately obvious to what extent we or future science-technology will ever be able to access such existences we call past. But in principle it does not seem to be altogether impossible; e.g. perhaps our local universe or zoo or region was produced (and is "recorded") by a Quasi-god (Super-person). (Moreover, according to Tandy 2007, past-directed time travel-viewing capacity is "likely" in the very-long-run.)

If we alter the Einsteinian Immortality paradigm so as to be an ontological (instead of physical-scientific) account of reality, then it seems to me that it must be true. Thus altered, let's call it **Ontological** Immortality. Accordingly, we ought to desire and seek physical-scientific theories that seem to lead to the ontological immortality account of reality. Current ("Einsteinian") scientific theory here gives the appearance of leaning in the ontologically desirable direction. Nevertheless, it's worth reminding ourselves that only truth can be the standard of truth! And, for finite beings, the wiser path to walk is one of proximate truths as distinguished from jumping wildly to (probably false) conclusions.

Spiritual Immortality

> Spiritual Immortality: When my body dies I will nevertheless continue to live; my spirit (as a multi-staged life or career) will continue to have experiences.

Accepting this view at face value without further alteration or embellishment seems difficult to me. Current experts tell us that the developmental order (in our local region or this local universe) was from energy to atoms to life to basic-mentality to human-mentality. The human person seems to be (both) body-and-mind (together) instead of a body (or a body with a mind) or a mind (or a mind with a body). Alternatively, one may be able to combine two or more other views of immortality to arrive at results roughly simulating this view.

Cosmic Immortality

> Cosmic Immortality: We came from the eternal cosmos and upon death will return to the cosmic mind.

Although I can see possible merit in this view, I'm not sure I am motivated to strongly defend it. My interest is not just in the immortality of all-reality (a necessary truth) or of cosmic mind but of my own personal immortality and the immortality of all persons. My attitude is that all-reality or that cosmic mind wants me, or should want me, to be interested in the immortality of all persons.

Physical Immortality

> Physical Immortality: The pattern of components that constitute my body (e.g. my brain) may be disrupted to the extent that the pattern no longer physically exists; restoration of the pattern to functional physical existence would resurrect me.

I am motivated by this view and it seems to be supported by the ontological immortality perspective I advocated above as certainly true. If we combine an ethical interest favoring the immortality of all persons with the ontological immortality paradigm, then we get (or so I think) an ethical or categorical imperative to develop scientific theories, technologies, and techniques for the ultimate purpose (sooner rather than later!) of physically resurrecting all persons no longer alive. Let's call this the **onto-resurrection imperative** (or, alternatively, our **common task**). Jacques Choron notes that: [10] "The main difficulty with personal immortality...is that once the naive position which took deathlessness and survival after death for granted was shattered, immortality had to be proved. All serious discussion of immortality became a search for arguments in its favor." "In order to be a satisfactory solution to the problems arising in connection with the fact of death, immortality must be first a 'personal' immortality, and secondly it must be a 'pleasant' one."

Entropy Is A Fake

Note that the "dismal" theory of thermodynamics in the form of its second law (the so-called "entropy" law) applies to closed systems. But given the context of part one above, we can now say: Gödel showed us that all-reality is **not** a closed system (see again endnote 6). "The entropy concept," according to Kenneth Boulding, "is an unfortunate one, something like phlogiston (which turned out to be negative oxygen), in the sense that entropy is negative potential. We can generalize the second law in the form of a law of diminishing potential rather than of increasing entropy, stated in the form: If anything happens, it is because there was a potential for it happening, and after it has happened that potential has been used up. This form of stating the law opens up the possibility that potential might be re-created..." [11] Again I emphasize that the second law does **not** really say that (all-reality's) potential is finite. Instead, let me suggest that the second law may be related to the arrow of time or to my statement above that "Once I do X instead of Y, X will **always** be the case."

Our Common Task: The Onto-Resurrection Project

Work beginning in the 20th century has laid the foundation for eventual realization of the onto-resurrection imperative. Developments have already taken us to the threshold of what has been called "practical time travel" – or what, loosely speaking, we may call "time travel". Once time travel becomes feasible in the 21st century, then we can proceed to more fully implement our common task of resurrecting all (rather than some) persons no longer alive. The first steps occurred in the 20th century on several fronts, including steps in the direction of suspended-animation, superfast-rocketry, and seg-communities. [12]

Experts tell us that the results of the population explosion (i.e. the size of the human population) will level off sometime in the 21st century (perhaps mid-century). Experts also tell us that current and ongoing industrial-technological activities are dangerously polluting our planet and causing global warming; global warming, in turn, can very easily lead to unprecedented injustices and upheavals in a terror-filled global-village of weapons of mass death and destruction. Presumably we should take global action against global dangers along the lines suggested by Al Gore, Jared Diamond, and other experts; see the Gore-related website about the practical generation of carbon-free electricity: <www.RepowerAmerica.org>; also see the Diamond-related website about "the world as a polder": <www.mindfu lly.org/Heritage/2003/Civilization-Collapse-EndJun03.htm>. But certainly too we can and should engage in additional terrestrial and extraterrestrial activities to prevent doomsday and improve the human condition. If we are not balanced and careful in our actions, myopia can provide us with badly-needed near-term clarity while preventing us from the broader vision required for survival, thrival, and the common task.

Terrestrial Implementation Of Our Common Task

Perfection of future-directed time travel in the form of suspended-animation (biostasis) seems feasible in the 21st century. I believe it even seems feasible to eventually offer it freely to all who want it. Jared Diamond has pointed out that: "If most of the world's 6 billion people today were in cryogenic storage and neither eating, breathing, nor metabolizing, that large population would cause no environmental problems." [13] Too, this might allow them to travel to an improved world in which they would be immortal. Since aging and all other diseases would have been conquered, they might not have to use time travel again unless they had an accident requiring future medical technology.

Extraterrestrial Implementation Of Our Common Task

But the onto-resurrection imperative demands more than immortality for those currently alive. In extraterrestrial space we can experiment (perhaps, for example, via Einsteinian or Gödelian past-directed time travel-viewing) with immortality for all persons no longer alive. Seg-communities (Self-sufficient Extra-terrestrial Green-habitats, or O'Neill communities) can assist us with our ordinary and terrestrial problems as well as assist us in completion of the onto-resurrection project. Indeed, in Al Gore's account of the global warming of our water planet, his parable of the frog is a central metaphor. Because the frog in the pot of water experiences only a gradual warming, the frog does not jump out. I add: Jumping off the water planet is now historically imperative; it seems unwise to put all of our eggs (futures) into one basket (biosphere).

I close with these words from Jacques Choron: "Only pleasant and personal immortality provides what still appears to many as the only effective defense against...death. But it is able to accomplish much more. It appeases the sorrow following the death of a loved one by opening up the possibility of a joyful reunion...It satisfies the sense of justice outraged by the premature deaths of people of great promise and talent, because only this kind of immortality offers the hope of fulfillment in another life. Finally, it offers an answer to the question of the ultimate meaning of life, particularly when death prompts the agonizing query [of Tolstoy], 'What is the purpose of this strife and struggle if, in the end, I shall disappear like a soap bubble?'" [14]

Summary

Above it was shown that mental-reality and all-reality are dimensions of reality which are not altogether reducible to any strictly physical-scientific paradigm. A more believable (general-ontological) paradigm was presented. Within this framework, the issue of personal immortality was considered. It was concluded that the immortality project, as a physical-scientific common-task to resurrect all dead persons, is ethically imperative. The imperative includes as first steps the development of suspended-animation, superfast-rocketry, and seg-communities.

Who Mary Is And What Mary Knows

> So this is who I am,
> and this is all I know.
> ....
> You are my only.
> ....
> We don't say goodbye.
> We don't say goodbye.
> With all my love for you.
> And what else may we do?
> We don't say goodbye.
>
> -- *Immortality* by the Bee Gees

Acknowledgements

I would like to thank Ser-Min Shei and his philosophy department at National Chung Cheng University (Taiwan) for their assistance.

I would like to thank the following for their comments on an earlier draft: Giorgio Baruchello, Ben Best, Tom Buford, Aubrey de Grey, William Grey, John Leslie, J. R. Lucas, Mike Perry, Edgar Swank, and Jim Yount.

Bibliography

Boulding (1981). Kenneth E. Boulding. *Ecodynamics: A New Theory of Societal Evolution*. Sage Publications: Beverly Hills. (First edition, 1978; this edition, 1981).

Bronowski (1965). Jacob Bronowski. *The Identity of Man*. Natural History Press: New York. See pages 78-80.

Bronowski (1966). Jacob Bronowski. "The Logic of Mind", *American Scientist*, 54 (1), March 1966, Pages 1-14.

Catterson (2003). Troy T. Catterson. "Letting The Dead Bury Their Own Dead: A Reply To Palle Yourgrau" in *Death And Anti-Death, Volume 1* edited by Charles Tandy. Ria University Press: Palo Alto, CA. (Pages 413-426).

Chaitin (1982). Gregory J. Chaitin. "Gödel's Theorem and Information", *International Journal of Theoretical Physics*, 21, [1982], Pages 941-954.

Choron (1973). Jacques Choron. "Death and Immortality" in Volume 1 (Pages 634-646) of *The Dictionary of the History of Ideas* edited by Philip P. Wiener. (1973=vols.1-4; 1974=index vol.). Charles Scribner's Sons: New York. Available at <http://etext.virginia.edu/DicHist/dict.html>.

Descartes (1637). René Descartes. *Discourse on the Method*. (Originally published anonymously in French, 1637). (Various translations available).

Diamond (2005). Jared Diamond. *Collapse: How Societies Choose to Fail or Succeed*. Viking: New York.

Fedorov (2008). Nikolai Fedorovich Fedorov. [Two websites about him:] <http://www.iep.utm.edu/f/fedorov.htm>; and, <http://www.quantium.plus.com/venturist/fyodorov.htm>.

Feigl (1958). Herbert Feigl. "The 'Mental' and the 'Physical'" in *Minnesota Studies in the Philosophy of Science: Volume II: Concepts, Theories, and the Mind-Body Problem* edited by Herbert Feigl, Michael Scriven, and Grover Maxwell. University of Minnesota Press: Minneapolis. (Pages 370-497). (See especially section "V.c." on pages 431-438).

Gödel (1931). Kurt Gödel. "Über Formal Unentscheidbare Sätze der *Principia Mathematica* und verwandter Systeme", Part I, *Monatschefte für Mathematik und Physik*, Volume XXXVIII, [1931], Pages 173-198. (Reprinted with English translation in *Kurt Gödel: Collected Works*, Volume 1, Oxford University Press: New York, 1986, Pages 144-195).

Gore (2006). Al Gore. *An Inconvenient Truth: The Planetary Emergency of Global Warming and What We Can Do About It*. Rodale Books: Emmaus, Pennsylvania. [This is the first book in history produced to offset 100% of the CO_2 emissions generated from production activities with renewable energy; this publication is carbon-neutral.]

Hartshorne (1962). Charles Hartshorne. *The Logic of Perfection*. Open Court: La Salle, Illinois.

Jackson (1982). Frank Jackson. "Epiphenomenal Qualia", *Philosophical Quarterly*, XXXII (32), April 1982, Pages 127-136.

Jackson (1986). Frank Jackson. "What Mary Didn't Know", *Journal of Philosophy*, LXXXIII (83), May 1986, Pages 291-295. Edgar Swank [ES] recently asked me this question: "What are we to do to avoid Mary's perception of color of her own body?" ES then suggested at least one possible answer: "Perhaps we need to keep Mary in a room illuminated only by a monochromatic (say red) light source. Then everything is a varying shade of the same color."

Kuhn (1962). Thomas Kuhn. *The Structure of Scientific Revolutions*. University of Chicago Press: Chicago. (Second edition enlarged, 1970).

Leslie (2007). John Leslie. *Immortality Defended*. Blackwell Publishing: Oxford.

Lucas (2008). J. R. Lucas. "[Section:] Gödelian Arguments" of his "Positive Logicality" chapter contribution in the present anthology. Or see "[Section:] Gödelian Arguments" at his <http://users.ox.ac.uk/~jrlucas/reasreal/reaschp6.pdf>. This Section is in chapter two of his book *Reason and Reality* (forthcoming in 2009 from Ria University Press.)

Nagel (1958). Ernest Nagel and James R. Newman. *Gödel's Proof*. Routledge: London. See pages 100-102. (First edition, 1958; this edition, 2002).

Nagel (1974). Thomas Nagel. "What Is It Like to Be a Bat?", *Philosophical Review*, LXXXIII (83), 4 (October 1974), Pages 435-450.

O'Neill (2000). Gerard K. O'Neill. *The High Frontier: Human Colonies in Space*. Apogee Books: Burlington, Ontario, Canada. (3rd edition). Also see: <http://www.space-frontier.org/HighFrontier/testimonial.html>.

Orwell (1949). George Orwell. *1984*. New American Library: New York. (First edition, 1949; this edition, 1961).

Penrose (1989). Roger Penrose. *The Emperor's New Mind*. Oxford University Press: New York.

Penrose (1990). Roger Penrose. "Précis", *Journal of Behavioral and Brain Sciences*, 13 (4), [1990], Pages 643-654.

Penrose (1994). Roger Penrose. *Shadows of the Mind*. Oxford University Press: New York.

Penrose (2005). Roger Penrose. *The Road to Reality: A Complete Guide to the Laws of the Universe*. Alfred A. Knopf: New York. (First edition, 2004; this edition, 2005).

Perry (2000). R. Michael Perry. *Forever For All: Moral Philosophy, Cryonics, And The Scientific Prospects For Immortality*. Universal Publishers: Parkland, FL.

Seg-communities (2008). [See these six websites about seg-communities (Self-sufficient Extra-terrestrial Green-habitats, or O'Neill communities):]
(1) <http://en.wikipedia.org/wiki/Space_colonization>;
(2) <http://www.nas.nasa.gov/About/Education/SpaceSettle ment>;
(3) <http://www.nss.org/settlement/space/index.html>;
(4) <http://www.segits.com>;
(5) <http://www.spaext.com>; and,
(6) <http://www.ssi.org>.

Strawson (1958). P. F. Strawson. "Persons" in *Minnesota Studies in the Philosophy of Science: Volume II: Concepts, Theories, and the Mind-Body Problem* edited by Herbert Feigl, Michael Scriven, and Grover Maxwell. University of Minnesota Press: Minneapolis. (Pages 330-353).

Strawson (1959). P. F. Strawson. *Individuals: An Essay in Descriptive Metaphysics*. Routledge: London. (1959, 1964, 1990).

Strawson (1985). P. F. Strawson. *Scepticism and Naturalism: Some Varieties*. Columbia University Press: New York.

Tandy (2007). Charles Tandy. "Types Of Time Machines And Practical Time Travel", *Journal Of Futures Studies*, 11(3), [February 2007], Pages 79-90.

Time-travel (2008). [See these websites about (1) time-travel; (2) suspended-animation; and, (3) superfast-rocketry:]
(1) <http://www.jfs.tku.edu.tw/11-3/A05.pdf>;
(2) <http://en.wikipedia.org/wiki/Greg_Fahy>; and,
(3) <http://en.wikipedia.org/wiki/Twin_paradox#Resolution_of_the_paradox_in_general_relativity>.
[Also see: Transhumanism (2008)].

Transhumanism (2008). [Two transhumanist websites:]
(1) <http://www.aleph.se/Trans>; and,
(2) <http://www.transhumanism.org>.

Endnotes

1. Feigl (1958), page 431.

2. Strawson (1985), chapter 3.

3. Strawson (1958), page 340.

4. Compare: Penrose (2005), chapters 1 and 34.

5. Descartes (1637), part IV.

6. Gödel (1931); Lucas (2008). J. R. Lucas has kindly suggested these additional references: Bronowski (1965); Bronowski (1966); Nagel (1958); Penrose (1989); Penrose (1990). Also see: Chaitin (1982).

7. Hartshorne (1962), page 88.

8. Quotation one: <http://www.iep.utm.edu/f/fedorov.htm>; Quotations two and three: "[Section] 3. On History" at <http://www.quantium.plus.com/venturist/fyodorov.htm>.

9. Compare: Leslie (2007), chapter 4.

10. Choron (1973), page 638. (Both quotations).

11. Boulding (1981), page 10.

12. Time-travel (2008); Seg-communities (2008).

13. Diamond (2005), page 494. This may be an exaggeration in that the production of liquid air/nitrogen requires energy; even so, Diamond would appear to be mostly correct here. But it is also conceivable that all or almost all power plants and related technologies will become carbon-neutral or even carbon-extracting. For example, see one of "Al Gore's websites" related to the practical generation of carbon-free electricity: <www.RepowerAmerica.org>. (Some environmentalists say that the additional step or capacity of carbon-extraction is required – or is at least desirable to make our lives easier. Whether practical carbon-extraction techniques would or would not require advanced molecular nanotechnology is not immediately obvious to me. Whether carbon-extraction, carbon-offsets, weather-modification, or terra-forming might be used as a doomsday weapon or weapon of mass death and destruction is yet another matter.)

14. Choron (1973), page 638.

CHAPTER TEN

The Future Of Scientific Simulations:
From Artificial Life To Artificial Cosmogenesis

Clément Vidal

■ Introduction

What will happen to the Earth and the Sun in the far future? The future story depicted by modern science is a gloomy one. In about 6 billion years, it will be the end of our solar system, with our Sun turning into a red giant star, making the surface of Earth much too hot for the continuation of life as we know it. The solution then appears to be easy: move. However, even if life would colonize other solar systems, there will be a progressive end of all stars in galaxies. Once stars have converted the available supply of hydrogen into heavier elements, new star formation will come to an end. In fact, the problem is worse. It is estimated that even very massive objects such as black holes will evaporate in about 10^{98} years (Adams and Laughlin 1997).

This scenario is commonly known as the "heat death", and says that the universe will irreversibly decay towards a state of maximum entropy **[b, d]**.

Letters and numbers in bold and brackets refer to the two maps in Annex 1; please consult this annex for more details.

If this "heat death" model is correct **[c]**, then it clearly means that the indefinite continuation of life is impossible in this universe **[f]**. What is the point of living in a universe doomed to annihilation? Ultimately, why should we try to solve mundane challenges of our daily lives and societies, if we can not even imagine a promising future for intelligent life in the universe? If we recognize this heat death **[1.12]**, then we should certainly do something to avoid it **[1.13]**, and thus try to change the future of the universe **[1.14]**.

A few authors have proposed some speculative solutions, but we will see that they are insufficient because none of them presently allows the indefinite continuation of intelligent life. We will instead argue that intelligent civilization will in the far future produce a new universe **[4.0]**. Although it sounds like a surprising proposition, resembling science fiction scenarios, we will consider it seriously and carefully.

It should be noted that the proposition of involving intelligent life into the fate of the universe is at odds with traditional science. Indeed, the modern scientific worldview has often suggested that the emergence of intelligence was an accident in a universe that is completely indifferent to human concerns, goals, and values (e.g. Weinberg 1993; Stenger 2007). I thus challenge this proposition, and another one that is commonly associated with it, which says that:

[a] intelligent civilization can not have a significant influence on cosmic evolution.

Our central focus is on the future of scientific simulations, and how important this activity could be in the far future, if intelligent civilization is to have influence on cosmic evolution. It is increasingly clear that simulations and computing resources are becoming main tools of scientific activity **[1.15]**. More concretely, at a smaller scale than the universe, we have already begun to produce and "play" with artificial worlds, with the practice of computer simulations. In particular, efforts in the Artificial Life (ALife) research field have shown that it is possible to create digital worlds with their own rules, depicting agents evolving in a complex manner. We will see that such simulations promise to become more and more complex and elaborate in the future.

In the first part ("Towards A Simulation Of An Entire Universe"), we argue that the path towards a simulation of an entire universe is an expected outcome of our scientific simulation endeavours. We then examine (in "Towards A Realization Of An Entire Universe") how such a simulation could be realized (instantiated, made physical) and solve the irreversible heat death of the universe, expected to happen at some future time.

■ Towards A Simulation Of An Entire Universe

In this section, we argue that simulating open-ended evolution not only in biology, but also to link it to physical evolution (a level below) and to cultural evolution (a level above) will be a long-term outcome of our scientific simulation endeavours. Such a simulation would allow us to probe what would happen if we would "replay the tape of the universe". We then discuss in more depth the status and potential usefulness of a simulation of an entire universe, making a distinction between *real-world* and *artificial-world* modelling. We outline and criticize the "simulation hypothesis", according to which our universe has been proposed to be just a simulation. Let us first summarize the historical trend of exponential increase of computing resources.

- **Increase of computing resources**

We may note two important transitions in the history of human culture. The first is the *externalization of memory* through the invention of writing. This allowed an accurate reproduction and a safeguard for knowledge. Indeed, knowledge could easily be lost and distorted in an oral tradition. The second is the *externalization of computation* through the invention of computing devices. The general purpose computer was inspired by the work of Church, Gödel, Kleene and Turing, and its formal specifications constitute the most general computing device (see Davis 2000 for a history of computation). The consequences of this last transition are arguably as significant -- or even more significant -- as the invention of writing. In particular, the changes induced by the introduction of computers in scientific inquiry are important, and remain underestimated and understudied; see however (Floridi 2003) for a good starting point.

Computing resources have grown exponentially, at least for over a century. There is much literature about this subject (see e.g. Kurzweil 1999; 2006). Moore's "law" famously states that the number of transistors doubles every 18 months on a single microprocessor **[1.21]**. Exponential increase in processing speed and memory capacity are direct consequences of the law. What are the limits of computer simulations in the future? Although there is no Moore's law for the efficiency of our algorithms, the steady growth in raw computational power provides free "computational energy" to increase the complexity of our simulations. This should lead to longer term and more precise predictions. Apart from the computational limitation theorems (uncomputability, the computational version of Gödel's theorem proved by Turing), the only limit to this trend is the physical limit of matter or the universe itself (Lloyd 2000; Krauss and Starkman 2004). As argued by Lloyd (2000; 2005) and Kurzweil (2006, 362) it should be noted that the ultimate computing device an intelligent civilization could use in the distant future is a very dense object, i.e. a black hole [1.22].

From a cosmic outlook, Moore's trend is in fact part of a much more general trend which started with the birth of galaxies. The cosmologist and complexity theorist Eric Chaisson proposed a quantitative metric to characterize the dynamic (not structural) complexity of physical, biological and cultural complex systems (Chaisson 2001; 2003). It is the *free energy rate density* (noted Φ_M) which is the rate at which free energy transits in a complex system of a given mass (Fig. 1). Its dimension is energy per time per mass (erg s^{-1} g^{-1}). Let us illustrate it with some examples (Chaisson 2003, p. 96). A star has a value ~1, planets ~10^2, plants ~10^3, humans ~10^4 and their brain ~10^5, current microprocessors ~10^{10}. According to this metric, complexity has risen at a rate faster than exponential in recent times [1.20]. We might add along this complexity increase, the hypothesis that there is a tendency to do ever more, requiring ever less energy, time and space; a phenomenon also called *ephemeralization* (Fuller 1969; Heylighen 2007), or "Space-Time Energy Matter" (STEM) compression (Smart 2008). This means that complex systems are increasingly localized in space, accelerated in time, and dense in energy and matter flows.

In Tomas Ray's simulation *Tierra* (Ray 1991), digital life competes for CPU time, which is analogous to energy in the organic world. The analogue of memory is the spatial resource. The agents thus compete for fundamental properties of computers (CPU time, memory) analogous to fundamental physical properties of our universe. This design is certainly one of the key reasons for the impressive growth of complexity observed in this simulation.

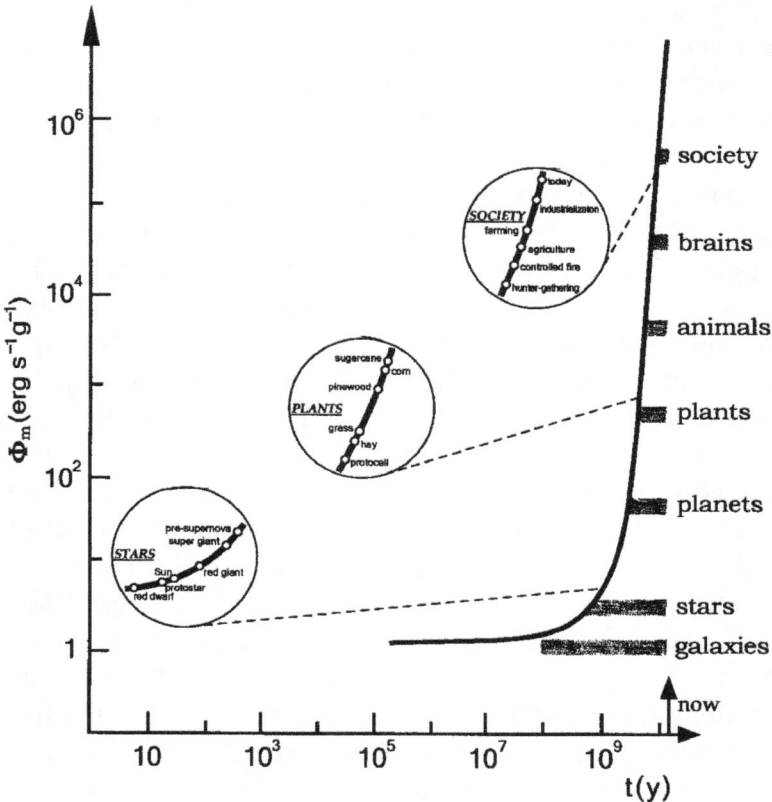

Fig. 1. Figure excerpted from (Chaisson 2003, p. 97). The original caption is: "The rise of free energy rate density, Φ_M, plotted as histograms starting at those times when various open structures emerged in Nature, has been rapid in the last few billion years, much as expected from both subjective intuition and objective thermodynamics. The solid curve approximates the increase in normalized energy flows best characterizing the order, form and structure for a range of systems throughout the history of the Universe. The circled

insets show greater detail of further measurements or calculations of the free energy rate density for three representative systems -- stars, plants and society -- typifying physical, biological and cultural evolution, respectively. Many more measures are found in Chaisson (2001)." Note that microprocessors are outside the scale of this diagram since they appear at 10^{10} on the Φ_M axis.

ZOOM IN on the circled insets (stars, plants, society):

STARS
Red dwarf
Protostar
Sun
Red giant
Super giant
Pre-supernova

PLANTS
Protocell
Hay
Grass
Pinewood
Corn
Sugarcane

SOCIETY
Hunter-gathering
Controlled fire
Agriculture
Farming
Industrialization
Today

- **Bridging physical, biological and cultural evolution**

We saw that a metric can be found to compare complex systems traditionally considered as different in nature. This important insight is just a first step towards bridging physical, biological and cultural evolution **[1.32]**. The information-theoretic endeavours are certainly going in this direction -- e.g. (Von Baeyer 2004; Prokopenko, Boschetti, and Ryan 2007; Gershenson 2007; Floridi 2003), as well as "Big History" thinkers such as (Christian 2004; Spier 2005).

Artificial Life (ALife) is a field of research examining systems related to life, its processes, and its evolution through simulations using either computer models (soft ALife), robotics (strong ALife), or biochemistry (wet ALife). A general challenge for ALife is to obtain an artificial system capable of generating open-ended evolution (Bedau et al. 2000). Some results have been obtained linking for example the evolution of language with quasi-biological traits (Steels and Belpaeme 2005). Working towards the design of a digital universe simulating the rise of levels of complexity in the physical, biological and cultural realms is the challenge of simulating an entire universe **[1.16]**. An important step in this direction, although it stays on the physical level, is the "Millennium Run" simulation, which starts from the very beginning of the universe to generate the large scale structures of the universe (Springel et al. 2005).

However, we must acknowledge important difficulties of conceptual, methodological and cultural integration between the different disciplines involved. In such an endeavour, human-made social and academic boundaries between disciplines of knowledge must be overcome **[1.31]**. I proposed to construct integrative scientific worldviews (or philosophies) with *systems theory*, *problem solving* and *evolutionary theory* as three generic interdisciplinary approaches (Vidal 2008). There should be a seamless link between simulations in physics, biology and social sciences (culture). If this would happen, we would have the basic tools to work towards a model and a simulation of the entire universe **[1.33; 2.0]**. In fact the search for such bridges is obviously necessary if we want to tackle such difficult problems as the origin of life, where we aim to explain the emergence of life out of physico-chemical processes.

- **Replaying the tape of the universe**

The biologist Stephen Jay Gould (1990) asked the famous question: "what would remain the same if the tape of life were replayed?". Paraphrasing and extending it to the universe, the question becomes: "what would remain the same if the tape of the universe were replayed?". We should first notice that the tape metaphor has its limits. Indeed, if the tape and its player were perfect, we should get exactly the same results when re-running the tape. Yet if our universe self-constructs, one question is whether small fluctuations could lead to slightly different outcomes, or very different ones if for example the system is chaotic.

By exploring other simulated universes, this approach would allow us to face one of the main difficulties in cosmology, which is that, as far as we know, there is only one object of study: our unique universe. More precisely, two fundamental limitations of current cosmology that Ellis (2005, sec. 3) has pointed out might then be addressed:

Thesis A1: The universe itself cannot be subjected to physical experimentation. *We cannot re-run the universe with the same or altered conditions to see what would happen if they were different, so we cannot carry out scientific experiments on the universe itself.* Furthermore,

Thesis A2: The universe cannot be observationally compared with other universes. *We cannot compare the universe with any similar object, nor can we test our hypotheses about it by observations determining statistical properties of a known class of physically existing universes.*

Interestingly, re-running the tape of the universe is also a very relevant research program for tackling the difficult "fine-tuning" problem in cosmology, which states that if any of a number of parameters, fundamental constants in physics and initial conditions in cosmology were slightly different, no complexity of any sort would come into existence; see e.g. (Leslie 1989) for a good review. To give just one example of fine tuning, let us consider the ratio of electrical and gravitational forces between protons, which is 10^{36}. Changes either in electromagnetism or in gravity « by only one part in 10^{40} would spell catastrophe for stars like the sun » (Davies 1984, p. 242).

Victor Stenger (1995; 2000) has performed a remarkable simulation of possible universes. He considered four fundamental constants, and then analysed "100 universes in which the values of the four parameters were generated randomly from a range five orders of magnitude above to five orders of magnitude below their values in our universe, that is, over a total range of ten orders of magnitude" (Stenger 2000). Anthony Aguirre did a similar work by exploring classes of cosmologies with different parameters (Aguirre 2001). These simulations are only an early attempt in simulating other possible universes, and the enterprise is certainly worth pursuing, with more complex models, more parameters to vary, etc.

The simulation of an entire universe can be seen as perhaps the ultimate challenge of simulations in science. But what kind of simulation would it be? What could it be used for?

To answer these questions we will now distinguish between two kinds of modelling.

▪ Real-world and artificial-world modelling

A computer simulation can be defined as a model where some aspects of the world are chosen to be modelled and the rest ignored. When in turn such a simplified model is run on hardware that is significantly more computationally efficient than the physical system being modelled, this makes it possible to run the model faster than the phenomena modelled, and thus to make predictions of our world. The paradigm of ALife strongly differs from traditional modelling, by studying not only "life-as-we-know-it", but also "life-as-it-could-be" (Langton 1992, sec. 1). We propose to extend this modelling technique to any *process* and not just to life, leading to the more general distinction of *processes-as-we-know-them* and *processes-as-they-could-be* (Red'ko 1999). We call the two kinds of modelling respectively *real-world* modelling and *artificial-world* modelling.

Real-world modelling is the endeavour to model *processes-as-we-know-them*. This includes traditional scientific modelling, such as models in physics, weather forecast models, but also applied evolutionary models, etc. The goal of such models is to better understand our world, and make predictions about it. For what would a *real-world* simulation of an entire universe be useful? At first glance, it would provide us very good understanding of and predictive power over our world. However, this view has some severe limitations. First, if the simulation is really of the entire universe, it should be "without anything left out". This is a strange situation, since it would imply that the model (simulation) is as complex as our universe. Such a simulation would thus not provide a way to systematically predict all aspects of our universe, because it would not be possible to run it faster than real physical processes. Another limiting argument is that more computational power does not necessarily mean better predictive abilities. This is pretty clear when considering chaotic systems such as the weather, which rapidly become unpredictable. A simulation still has to be simpler than reality if it is to be of any practical use. This means that in the context of "replaying the tape of our universe", we would still have to investigate a simplified simulation of our universe.

Artificial-world modelling is the endeavour to model *processes-as-they-could-be*. The formal fundamental rules of the system (of life in the case of ALife) are sought. The goal of ALife is not to model life exactly as we know it, but to decipher the most simple and general principles underlying life and to implement them in a simulation. With this approach, one can explore new, different life-like systems. Stephen Wolfram (2002) has a similar approach by exploring different rules and initial conditions on cellular automata, and observing the resulting behaviour of the system. It is legitimate to emphasize that this is a "new kind of science". Indeed, this is in sharp contrast with traditional science focusing on modelling or simulating reality. There is thus a creative aspect in the artificial-world modelling, which is why many artists have enthusiastically depicted imaginary ALife worlds. For what would an artificial-world simulation of an entire universe be useful? We would be able not only to "replay the tape of *our* universe", but also to play and replay the tape of *other possible* universes (thus tackling limitations A1 and A2 explicated by Ellis) **[2.1; 2.2]**. We saw that simulation constitutes a research program for tackling the fine-tuning issue in cosmology **[2.3]**. The concept of "a universe" then needs to be redefined and extended, since we only know by definition our unique universe.

Should this artificial world modelling of an entire universe be interpreted as a *simulation* or as a *realization* (Pattee 1989)? To start, let us consider the first possibility, with the *simulation hypothesis.*

- **The simulation hypothesis**

Let us assume what we have argued in the previous section, i.e. that intelligent life will indeed be able at some point to simulate an entire universe. If such a simulation is purely digital, thus pursuing the research program of soft ALife, this leads to the *simulation hypothesis*, which has two main aspects. First, looking into the future, it means that we would effectively create a whole universe simulation, as has been imagined in science fiction stories and novels such as the ones of Isaac Asimov (1956) or Greg Egan (2002). Very well then! A second possibility is that we ourselves could be part of a simulation run by a superior intelligence (e.g. Bostrom 2003; Barrow 2007; Martin 2006). Although these scenarios are fascinating, they suffer from two fundamentals problems. First, the "hardware problem": on what physical device would such a simulation run? Is there an infinity of simulation levels? Second, such an hypothesis violates Leibniz' logical principle of the identity of the indiscernibles. Leibniz' principle states that "if, for every property F, object x has F if and only if object y has F, then x is identical to y". Let x be reality, and y be the supposed simulated universe we would be living in. If we have no way to distinguish between them, they are identical. Unless we find a "bug" in reality, or a property F that could only exist in a simulation and not in reality, this hypothesis seems useless. A more comprehensive criticism of these discussions can be found in (McCabe 2005).

The ontological status of this simulation would be reflected by the states of the hardware running it, whatever the realistic nature of the simulation. From this point of view, we can argue that it remains a *simulation*, and not a *realization* (Harnad 1994). Is there another possibility for realizing the simulation of an entire universe? That is what we will explore now.

■ Towards A Realization Of An Entire Universe

We first outline some aspects concerning the far future of the universe. We then put forward a philosophical approach to tackle this problem, and outline a speculative solution called "artificial cosmogenesis".

▪ The heat death problem

We outlined in introduction the heat death problem. Consider the second law of thermodynamics which is one of the most general laws of physics. It states that the entropy of an isolated system will tend to increase over time. Hermann von Helmholtz applied it to the universe as a whole in 1854 to state the heat death problem, i.e. that the universe will irreversibly go towards a state of maximum entropy. Modern cosmology shows that there are some other models of the end of the universe -- such as Big Bounce, Big Rip, Big Crunch...; see (Vaas 2006) for an up-to-date review. The point is that none of them allows the possibility of the indefinite continuation of life as we know it. The study of the end state of the universe, or *physical eschatology*, is a scattered but exciting field of research that we cannot detail more here; see (Ćirković 2003) for an extensive literature guide.

Some speculative scenarios have been proposed to tackle this problem. They all suppose as we do in this paper that "intelligent civilization *can* have significant influence on cosmic evolution" **[4.1]**; but also that in the future, life will be very different from the one we know. Let us mention some of them. Dyson proposed that life and communication can continue "forever", utilizing a finite store of energy (Dyson 1979); the "final anthropic principle" put forward by Barrow and Tipler (1986) proposes that intelligent information-processing will never die out. Interestingly, under certain conditions, it is theoretically possible to make computing a reversible process (Bennett 1982; Landauer 1991; Krauss and Starkman 2000). If we could make this happen, this might be a way to possibly have "life" continue for an indefinite amount of time.

These speculations are remarkable in the sense that they attempt to find ways for intelligent life to survive forever. However, they assume the additional hypothesis that life should take another "information-like" form. Krauss and Strakman (2000) showed that there are serious difficulties to the scenario proposed by Dyson. The reversible computation scenario is also not sustainable in the long run, since, as Krauss and Strakman argue, no finite system can perform an infinite number of computations with a finite amount of energy. Furthermore, these scenarios give no clear link to the increasing abilities of intelligent life to model the universe, nor do they relate to the fine-tuning problem.

In an optimistic picture, that is if our civilization does not self-destruct (or if it does, we can add the hypothesis that we are not alone in the universe...), we can see the heat death problem as the longest-term problem for intelligent life in the universe. How should we react to it? Charles Darwin's thought remains perfectly relevant: "Believing as I do that man in the distant future will be a far more perfect creature than he now is, it is an intolerable thought that he and all other sentient beings are doomed to complete annihilation after such long-continued slow progress" (Darwin 1887, p. 70).

▪ A philosophical approach for a speculative topic

Before proposing another possible solution to the HEAT DEATH problem, we have to make a methodological clarification. The solution proposed in the next section will be approached from a *speculative* philosophical stance, as opposed to *critical* philosophy (Broad 1924). I have proposed a general philosophical framework to rationally construct speculative theories: (Vidal 2007). We should be well aware of the difficulty of the question we are tackling; an age-old philosophical problem which is: "what is the ultimate fate of humanity and the universe in the very distant future?". This problem is philosophical because (1) we do not have unambiguous empirical or experimental support to favour a unique outcome; and, (2) it is such an ambitious question, that the proposed answer can only be tentative and speculative. It is however still very worth considering because the philosophical inquiry aims to advance our most profound questions here and now, whatever their difficulty and our limited knowledge.

▪ Artificial cosmogenesis

The cosmologist Lee Smolin proposed a theory called Cosmological Natural Selection (CNS) in order to tackle the fine-tuning problem (Smolin 1992; 1997). According to this natural selection of universes theory, black holes give birth to new universes by producing the equivalent of a Big Bang, which produces a baby universe with slightly different physical properties (constants, laws). This introduces variation, while the differential success in self-reproduction of universes via their black holes provides the equivalent of natural selection. This leads to a Darwinian evolution of universes whose properties are fine tuned for black hole generation, a prediction that can in principle be falsified.

Smolin is not the only cosmologist reasoning with multiple universes, comprising an extended ensemble called a multiverse. Although the idea of a multiverse is a speculative one, it is increasingly popular among many cosmologists. New universes are generally theorized to appear from the inside of black holes, or from the Big Bang itself [3.0; 3.1]. However, Kuhn (2007) distinguished many kinds of other multiverse models: by disconnected regions (spatial); by cycles (temporal); by sequential selection (temporal); by string theory (with minuscule extra dimensions); by large extra dimensions; by quantum branching or selection; by mathematics and even by all possibilities, whatever this may mean. Among these multiverse theories, Smolin's CNS is arguably the most scientifically testable (Smolin 2007).

It should be noted however that in Smolin's theory, (1) the roles of life and intelligence in the universe are incidental, as they are in the modern scientific worldview. Which is, let us remember, the main assumption we challenge here. Another problem is that (2) the theory does not propose a specific mechanism for the variation of universe parameters beyond the assumption of randomness. Is it possible to overcome these two shortcomings? A few authors have dared to extend CNS by including intelligent life into this picture, correcting those two problems and also bringing indirectly a possible solution to the HEAT DEATH problem (Crane 1994; Harrison 1995; Baláz 2005; Gardner 2000; 2003; Smart 2008). Simply stated, the thesis is that advanced intelligent civilization will solve the HEAT DEATH problem by reproducing the universe. This direction can be seen as the ultimate challenge of strong/wet ALife, to realize a new universe.

Let us note however that there is not (yet) a uniform terminology among the five mentioned authors. Inspired by Smolin's terminology we could speak of a "Cosmological

Artificial Selection" (CAS), artificial selection on simulated universes enhancing natural selection of real universes (Barrow 2001, 151). The biological analogy is interesting here. Humans who practice artificial selection on animals do not "design" or "create" new organisms, nor do they replace natural selection. They just try to foster some traits over others. In CNS, many generations of universes are needed to randomly generate a fine tuned complex universe. In CAS, the extensive simulations prior to the replication event would presumably help greatly to generate a fine tuned universe, a universe that is robust to complexity emergence.

We do not attempt here to go into the details of how a CAS could become realized. However, let us first remember that black holes might be the ultimate computing device for an intelligent civilization [1.22], and that they constitute a possible gateway for the emergence of a new universe in some multiverse theories [3.1].

Along with ALife, constituting an artificial biology, Pattee (1989) suggested to also consider an "artificial physics". As in ALife, this "Artificial Cosmogenesis" discipline would have two parts. One focusing on "software" universe simulations using computer models (analogous to soft ALife); the other focusing on implementing the software in reality (analogous to strong/wet ALife). It is clear however that the analogue of soft ALife (universe simulation) is only in its infancy, and the analogue of strong/wet ALife (universe realization) lies in the far future.

This solution to the heat death problem gives a general challenge to intelligence in the universe: to continue to explore and understand the functioning of our universe so as to possibly reproduce it in the far future [2.3; 4.0]. This would make the indefinite continuation of life possible, yet in another universe [4.2]. This scenario also fits with the ultimate goal of evolution as a whole: survival. It is likely to be a difficult and stimulating enough challenge to encourage and occupy intelligent civilization for the foreseeable future.

The degree of control that intelligence could have in this process still has to be discovered. For example, how much might the physical properties of our existing universe (physics of black holes, etc.) constrain the realization of a new universe? Furthermore, the issue of the ethical responsibility of humanity in this proposition is outside the scope of this paper and remains to be explored; see however (Gardner 2003, Part 6) and (Smart 2008) for two different viewpoints.

■ Conclusion

The use of scientific simulations has constituted a revolution in the way we practice science. We have outlined the fast-moving changes occurring in our universe, and argued that the limit of scientific simulations is the simulation of an entire universe. Furthermore, we have formulated an hypothesis that the heat death of complexity in our universe could be avoided through an *artificial cosmogenesis,* a discipline analogous to artificial life.

Scientific inquiry today undertakes to understand our world; in the future, this will be increasingly aided by simulations of our and other possible universes. Such simulations would be indispensable tools if intelligent civilization moves towards an artificial cosmogenesis.

■ Annex 1: Logical Structure Of The Paper

This annex presents the logical structure of the main arguments in this paper -- represented by two maps. The problem is mapped in Fig. 2 and the proposed solution in Fig. 3 – Fig. 4. For an easier back-and-forth between the paper and the maps, the blocks are lettered/numbered in the map (letters for Fig. 2; numbers for Fig. 3 – Fig. 4) and those letters/numbers appear in bold and brackets in the text.

This approach provides an *externalization of reasoning* so that arguments can be clearly visualized. This brings many benefits, such as:

- Allowing the reader to quickly and clearly grasp the logic of the argumentation.
- Presenting an alternative structure of the content of the paper. The table of contents and the abstract tend to present a rhetorical (and not logical) structure.
- Allowing the possibility of a constructive discussion of assumptions and deductions. For example, a critique can say "the core problem is not P but Q"; or "I disagree that hypothesis [X.XX] leads to [Y.YY], you need implicit hypothesis Z, ..." or "hypothesis [Z.ZZ] is wrong because"; or "there is another solution to your problem, which is..." etc.

It should be clear however that reading those maps cannot replace the reading of the paper. Only the core reasoning is mapped, sometimes even in a simplified way.

To draw those maps, we used some of the insights of Eliyahu Goldratt's Theory of Constraints (TOC) and its "Thinking Process" (see Goldratt and Cox 1984; Goldratt Institute 2001; Scheinkopf 1999). The TOC is a well proven management technique widely used in finance, distribution, project management, people management, strategy, sales and marketing. We see it and use it as part of a generic problem solving toolbox, where causes and effects are mapped in a transparent way. In our paper, the core problem is: "how to make the indefinite continuation of life possible?"; and the proposed solution is that "intelligent civilization can reproduce the universe".

In this TOC framework three fundamental questions are employed to tackle a problem:

(1) What to change?

A core problem is identified as the *undesirable effect*, and mapped in a "Current Reality Tree" (CRT); see Fig. 2.

(2) To what to change?

A solution is proposed and mapped in a "Future Reality Tree" (FRT), which leads to the *desirable effect*; see Fig. 3 – Fig. 4.

(3) How to cause the change?

A plan is developed to change from CRT to FRT. This third step in the context of this paper is even more speculative, so it is almost not developed and thus not mapped. To tackle the problem in practice, six important questions should be addressed, constituting the "six layers of resistance to change". These questions can be used to trigger discussions (Goldratt Institute 2001, p. 6):

 (1) Has the right problem been identified?

 (2) Is this solution leading us in the right direction?

 (3) Will the solution really solve the problems?

 (4) What could go wrong with the solution? Are there any negative side-effects?

 (5) Is this solution implementable?

 (6) Are we all really up to this?

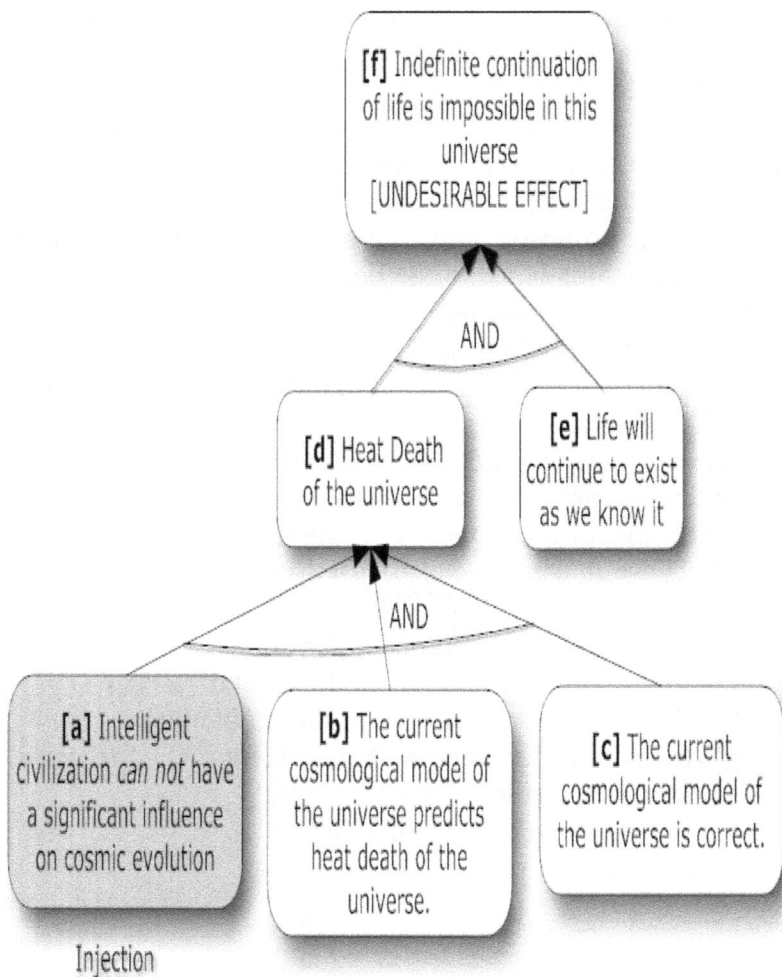

Fig. 2. The Current Reality Tree (CRT) represents the core problem underneath this paper (how to make the indefinite continuation of life possible?). The "injection" (grayed) is the proposition which is challenged. It is the statement that [a]: "intelligent civilization can not have a significant influence on cosmic evolution".

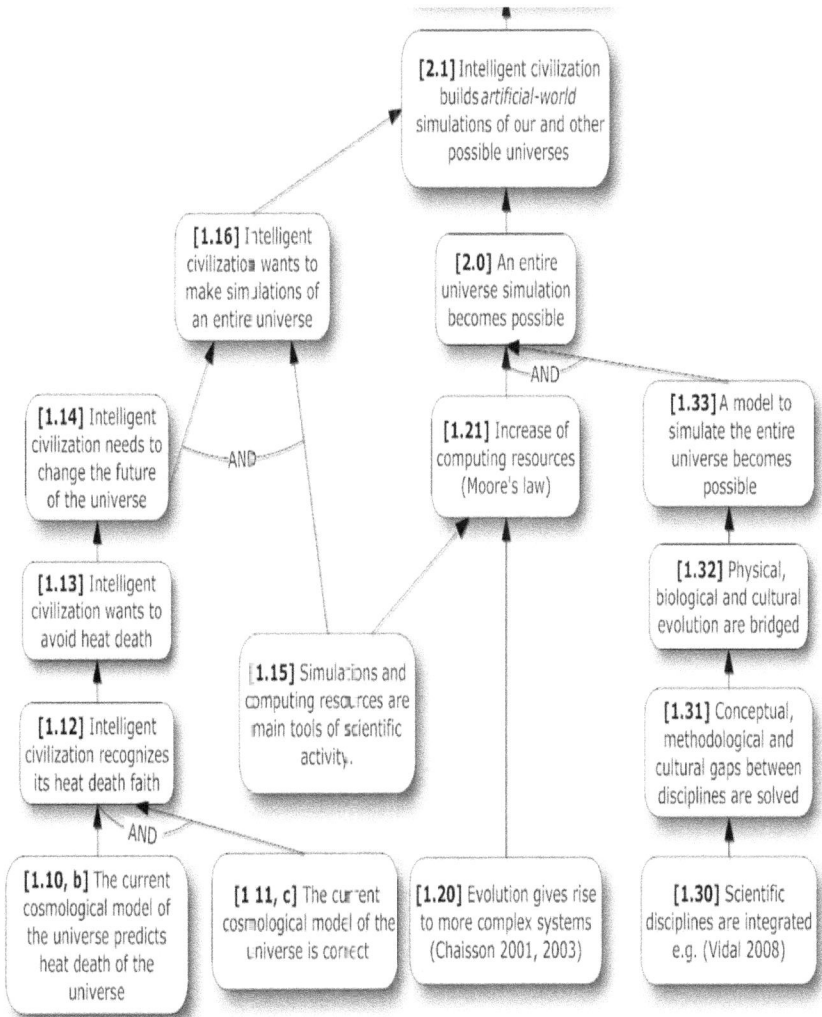

Fig. 3. The Future Reality Tree (FRT) shows the proposed solution to the problem mapped in the CRT (Fig. 2). The diagram can be read by increasing numerical order. See Fig. 4 (next page) for the continuation of the tree after **[2.1]**.

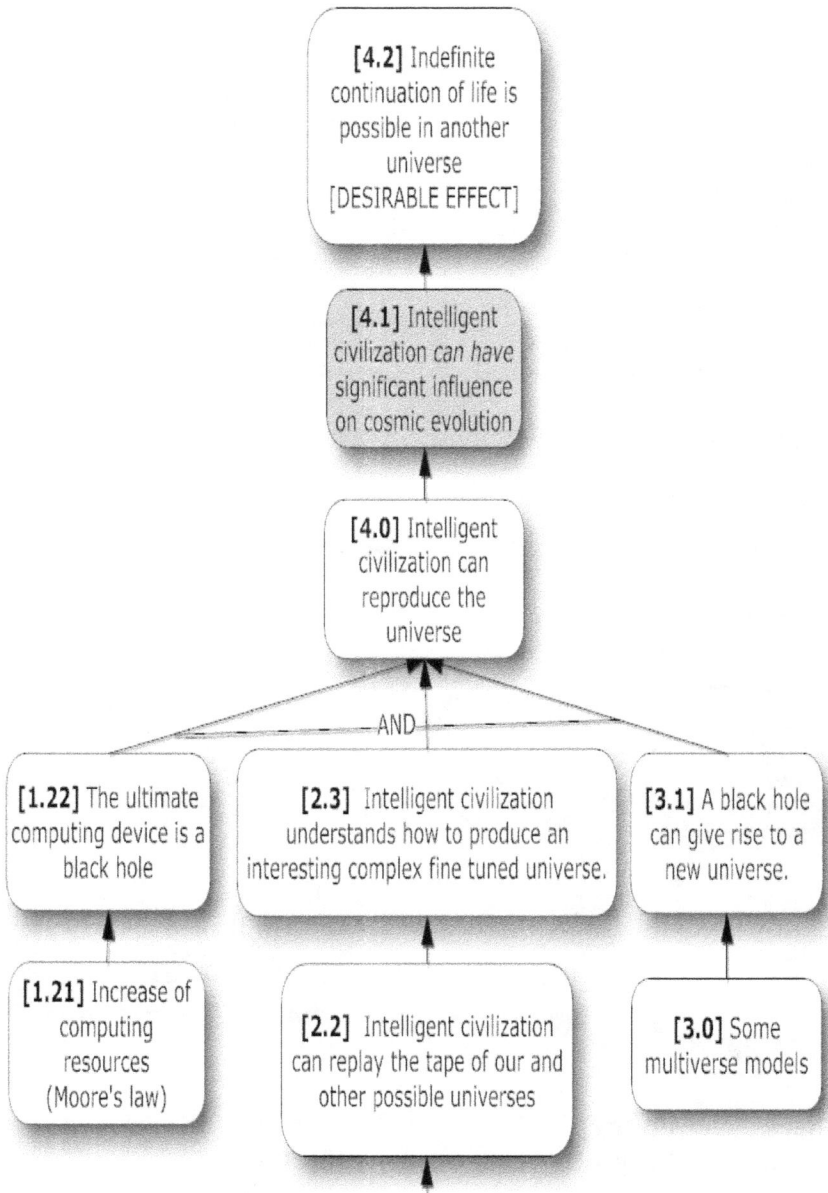

Fig. 4. Continuation of the FRT (Future Reality Tree) of Fig. 3. The "injection" chosen to solve the core problem is **[4.1]**: "intelligent civilization can have significant influence on cosmic evolution".

Acknowledgements

I thank three anonymous referees for their sharp comments, my colleagues Milan Ćirković, Carlos Gershenson, Francis Heylighen, Mark Martin, Marko Rodriguez and John Smart for their rich and valuable feedback. I thank Eric Chaisson for his comments and his kind permission to reproduce his curve (Fig. 1). I especially thank Piet Holbrouck for introducing me to the theory of constraints, which gave rise to the annex of this paper. Researchers interested in the topics of this paper (and others) are welcome to join the Evo Devo Universe research community at www.evodevouniverse.com.

References

Adams, F. C., and G. Laughlin. 1997. A Dying Universe: The Long-Term Fate and Evolution of Astrophysical Objects. *Reviews of Modern Physics* 69, no. 2: 337-372. http://arxiv.org/abs/astro-ph/9701131.

Aguirre, Anthony. 2001. The Cold Big-Bang Cosmology as a Counter-example to Several Anthropic Arguments. *astro-ph/0106143* (June 7). http://arxiv.org/abs/astro-ph/0106143.

Asimov, I. 1956. The Last Question. *Science Fiction Quarterly* 12, no. 4: 715. http://www.multivax.com/last_question.html.

Baláz, BA. 2005. The Cosmological Replication Cycle, the Extraterrestrial Paradigm and the Final Anthropic Principle. *Diotima*, no. 33: 44-53. http://astro.elte.hu/~bab/seti/IACP12z.htm.

Barrow, J. D. 2001. Cosmology, Life, and the Anthropic Principle. *Annals of the New York Academy of Sciences* 950, no. 1: 139.

---. 2007. Living in a Simulated Universe. In *Universe of Multiverse?*, ed. B. Carr, 481-486.

Barrow, J. D., and F. J. Tipler. 1986. *The Anthropic Cosmological Principle*. Oxford University Press.

Bedau, M. A., J. S. McCaskill, N. H. Packard, et al. 2000. Open Problems in Artificial Life. *Artificial Life* 6, no. 4: 363-376. http://people.reed.edu/~mab/publications/papers/ALife-6-4.pdf.

Bennett, C. H. 1982. The thermodynamics of computation: a review. *International Journal of Theoretical Physics* 21, no. 12: 905-940.

Bostrom, N. 2003. Are You Living in a Computer Simulation? *Philosophical Quarterly* 53, no. 211: 243-255. http://www.simulation-argument.com/simulation.pdf.

Broad, C. D. 1924. Critical and Speculative Philosophy. *Contemporary British philosophy: Personal statements*: 75-100. http://www.ditext.com/broad/csp.html.

Chaisson, E. J. 2001. *Cosmic Evolution: The Rise of Complexity in Nature*. Harvard University Press.

---. 2003. A Unifying Concept for Astrobiology. *International Journal of Astrobiology* 2, no. 02: 91-101. http://www.tufts.edu/as/wright_center/eric/reprints/unifying_concept_astrobio.pdf.

Christian, D. 2004. *Maps of Time: An Introduction to Big History*. University of California Press.

Ćirković, M. M. 2003. Resource Letter: PEs-1: Physical eschatology. *American Journal of Physics* 71: 122. http://www.aob.bg.ac.yu/~mcirkovic/Cirkovic03_RESOURCE_LETTER.pdf.

Crane, L. 1994. Possible Implications of the Quantum Theory of Gravity: An Introduction to the Meduso-Anthropic Principle. http://arxiv.org/abs/hep-th/9402104.

Davies, P.C.W. 1984. *Superforce the search for a grand unified theory of nature*. New York: Simon and Schuster.

Davis, M. 2000. *The Universal Computer: The Road from Leibniz to Turing*. W. W. Norton & Company, October.

Dyson, F. 1979. Time Without End: Physics and Biology in an Open Universe. *Review of Modern Physics* 51: 447-460. http://www.think-aboutit.com/Misc/time_without_end.htm.

Egan, G. 2002. *Schild's Ladder*. Eos.

Ellis, G. F. R. 2005. Philosophy of Cosmology. In *Handbook in Philosophy of Physics*, ed. J Butterfield and J Earman, 1183-1285. Elsevier. http://arxiv.org/abs/astro-ph/0602280.

Floridi, L., ed. 2003. *The Blackwell Guide to the Philosophy of Computing and Information*. Ed. L. Floridi. Blackwell Publishing.

Fuller, R. B. 1969. *Utopia Or Oblivion: The Prospects for Humanity*. Overlook Press.

Gardner, J. N. 2000. The Selfish Biocosm: complexity as cosmology. *Complexity* 5, no. 3: 34-45.

---. 2003. *Biocosm. The New Scientific Theory of Evolution: Intelligent Life is the Architect of the Universe*. Inner Ocean Publishing.

Gershenson, C. 2007. The World as Evolving Information. In *Proceedings of Seventh International Conference on Complex Systems ICCS*, ed. Y. Bar-Yam. http://uk.arxiv.org/abs/0704.0304.

Goldratt Institute. 2001. Theory of Constraints and its Thinking Processes - A Brief Introduction . Whitepaper. http://www.goldratt.com/toctpwhitepaper.pdf.

Goldratt, E. M., and J. Cox. 1984. *The Goal: A Process of Ongoing Improvement*. 3rd ed. Great Barrington, MA: North River Press.

Gould, S. J. 1990. *Wonderful Life: The Burgess Shale and the Nature of History*. WW Norton & Company.

Harnad, S. 1994. Levels of Functional Equivalence in Reverse Bioengineering: The Darwinian Turing Test for Artificial Life. *Artificial Life* 1, no. 3: 293-301. http://eprints.ecs.soton.ac.uk/archive/00003363/02/ha rnad94.artlife2.html.

Harrison, E. R. 1995. The Natural Selection of Universes Containing Intelligent Life. *Quarterly Journal of the Royal Astronomical Society* 36, no. 3: 193-203. http://adsabs.harvard.edu/full/1996QJRAS..37..369B.

Heylighen, F. 2007. Accelerating Socio-Technological Evolution: from ephemeralization and stigmergy to the global brain. In *Globalization as an Evolutionary Process: Modeling Global Change*, ed. George Modelski, Tessaleno Devezas, and William Thompson, 286-335. Routledge. London . http://pespmc1.vub.ac.be/Papers/AcceleratingEvoluti on.pdf.

Krauss, L. M., and G. D. Starkman. 2000. Life, the Universe, and Nothing: Life and Death in an Ever-expanding Universe. *The Astrophysical Journal* 531, no. 1: 22-30. http://arxiv.org/pdf/astro-ph/9902189.

---. 2004. Universal Limits on Computation. http://arxiv.org/abs/astro-ph/0404510.

Kuhn, R. L. 2007. Why This Universe? Toward a Taxonomy of Possible Explanations. *Skeptic*.

Kurzweil, R. 1999. *Age of Spiritual Machines: When Computers Exceed Human Intelligence*. Penguin USA New York, NY, USA.

---. 2006. *The Singularity Is Near: When Humans Transcend Biology*. Penguin Books.

Landauer, R. 1991. Information is physical. *Physics Today* 44, no. 5: 23-29.

Langton, C. G. 1992. *Artificial Life*. Addison-Wesley. http://www.probelog.com/texts/Langton_al.pdf.

Leslie, J. 1989. *Universes*. Routledge.

Lloyd, S. 2000. Ultimate Physical Limits to Computation. *Nature* 406: 1047-1054. http://www.hep.princeton.edu/~mcdonald/examples/ QM/lloyd_nature_406_1047_00.pdf.

---. 2005. *Programming the Universe: A Quantum Computer Scientist Takes on the Cosmos*. Vintage Books.

Martin, M. 2006. *The Mocking Memes: A Basis for Automated Intelligence*. AuthorHouse. Published under pseudonym Evan Louis Sheehan, October 11. http://evanlouissheehan.home.comcast.net/~evanlouis sheehan/TheMockingMemes.pdf.

McCabe, G. 2005. Universe Creation on a Computer. *Studies In History and Philosophy of Science Part B: Studies In History and Philosophy of Modern Physics* 36, no. 4 (December): 591-625. doi:10.1016/j.shpsb.2005.04.002. http://philsci-archive.pitt.edu/archive/00001891/01/UniverseCreati onComputer.pdf.

Pattee, H. H. 1989. Simulations, Realizations, and Theories of Life. *Artificial Life* 6: 63-78.

Prokopenko, M., F. Boschetti, and A. J. Ryan. 2007. An Information-Theoretic Primer on Complexity, Self-Organisation and Emergence. *Advances in Complex Systems*. http://www.worldscinet.com/acs/editorial/paper/5183 631.pdf.

Ray, T. S. 1991. An Approach to the Synthesis of Life. *Artificial Life II* 10: 371–408.

Red'ko, V. G. 1999. Mathematical Modeling of Evolution. *in: F. Heylighen, C. Joslyn and V. Turchin (editors): Principia Cybernetica Web (Principia Cybernetica, Brussels)*. http://pespmc1.vub.ac.be/MATHME.html.

Scheinkopf, L. J. 1999. *Thinking for a Change: Putting the Toc Thinking Processes to Use*. CRC Press.

Smart, J. 2008. Evo Devo Universe? A Framework for Speculations on Cosmic Culture. In *Cosmos and Culture*, ed. S. J. Dick. To appear. http://accelerating.org/downloads/SmartEvoDevoUni v2008.pdf.

Smolin, L. 1992. Did the Universe evolve? *Classical and Quantum Gravity* 9, no. 1: 173-191.

---. 1997. *The Life of the Cosmos*. Oxford University Press, USA.

---. 2007. Scientific Alternatives to the Anthropic Principle. In *Universe of Multiverse?*, ed. B. Carr, 323-366. Cambridge University Press.

Spier, F. 2005. How Big History Works: Energy Flows and the Rise and Demise of Complexity. *Social History and Evolution* 1, no. 1.

Springel, V., S. D. M. White, A. Jenkins, et al. 2005. Simulations of the Formation, Evolution and Clustering of Galaxies and Quasars. *Nature* 435: 629-636. http://astronomy.sussex.ac.uk/~petert/archive/svirgo05.pdf.

Steels, L., and T. Belpaeme. 2005. Coordinating Perceptually Grounded Categories Through Language: A Case Study for Colour. *Behavioral and Brain Sciences* 28, no. 04: 469-489.

Stenger, V. J. 1995. *The Unconscious Quantum: Metaphysics in Modern Physics and Cosmology.* Prometheus Books.

---. 2000. Natural Explanations for the Anthropic Coincidences. *Philo* 3, no. 2: 50-67.

Vaas, R. 2006. Dark Energy and Life's Ultimate Future. In *The Future of Life and the Future of our Civilization*, ed. V. Burdyuzha and G. Kohzin, 231-247. Dordrecht : Springer. http://arxiv.org/abs/physics/0703183.

Vidal, C. 2007. An Enduring Philosophical Agenda. Worldview Construction as a Philosophical Method. *Submitted for publication.* http://cogprints.org/6048/.

---. 2008. What is a worldview? Published in Dutch as: "Wat is een wereldbeeld?". In *Nieuwheid denken. De wetenschappen en het creatieve aspect van de werkelijkheid*, ed. Hubert Van Belle and Jan Van der Veken. Leuven: In press, Acco. http://cogprints.org/6094/.

Von Baeyer, H. C. 2004. *Information: The New Language of Science*. Harvard University Press.

Wolfram, S. 2002. *A New Kind of Science*. Wolfram Media Inc., Champaign, IL.

Death And Anti-Death Series By Ria University Press

INDEX

Death And Anti-Death Terminology

R. Michael Perry

H

I

K

L

Lucas, J. R., 8, 23, 29, 157, 220, 221, 277, 279, 282

M

Mind, 11, 18, 28, 31, 40, 46, 51, 55, 60, 65, 78, 101, 102, 104, 105, 109, 110, 127, 141, 142, 143, 144, 145, 146, 147, 154, 155, 188, 202, 203, 205, 244, 246, 254, 256, 269, 271, 272
Monetary policy, 27
Motivation, 27, 40, 223

N

Neoplatonism, 144

O

Ontological argument, 28, 53, 55, 57, 58, 78, 142
Ontology, 31, 264, 269

P

Panpsychism, 28, 56, 61
Paradigm, 31, 37, 39, 40, 168, 170, 207, 264, 266, 267, 268, 270, 272, 276, 294
Penrose, Roger, 8, 24, 30, 222, 238, 239, 240, 241, 280, 282
Perry, R. Michael, 8, 24, 30, 262, 263, 277, 280
Personal growth, 31, 244, 246, 247
Post-humanism, 28
Problem of dilution, 31, 245, 248, 251, 252, 259
Problem of stagnation, 31, 244, 249
Process philosophy, 28
Psychoactive drugs, 28, 101, 102, 103, 105, 106, 107, 108, 111, 114, 115, 116, 117, 122, 123

U

Universes, 29, 127, 131, 135, 138, 139, 140, 141, 145, 146, 153, 154, 156, 245, 292, 293, 295, 299, 300, 301, 302

V

Verstehen (understanding), 30, 203, 205

W

Wellbeing, 27, 34, 39, 110
Weyl curvature, 30, 229

ABOUT THE EDITOR

Dr. Charles Tandy received his Ph.D. in Philosophy of Education from the University of Missouri at Columbia (USA) before becoming a Visiting Scholar in the Philosophy Department at Stanford University (USA). He is author or editor of numerous publications, including the ***Death And Anti-Death*** series of anthologies from Ria University Press. Dr. Tandy, along with Nobel Laureates and others, is a member of the Board of Advisors of the Lifeboat Foundation. Indeed, Dr. Tandy is a board member, advisor, or consultant to a variety of charitable and educational institutions.

Dr. Tandy's research and teaching in the general humanities (liberal arts) typically relates to one or more of the following four areas: Biomedical Ethics; Futuristic Studies; Global History; Interdisciplinary Philosophy. Much of Dr. Tandy's work focuses on "big picture" issues related to the future of humanity. For example, how may our worldviews, technologies, and decisions affect future human existence? (Such profoundly important issues are under-explored and under-funded. Too often, such studies are flawed by discipline-centric, ethno-centric, or present-centric thinking. In life, politics, and scholarship, our interdisciplinary, global, and foresighted perspectives are dangerously underdeveloped.)

Dr. Tandy is a pioneer in time travel and suspended animation. He is author of scholarly publications on the ethics and metaphysics of time travel and suspended animation. He has argued that sooner or later the perfection of: (1) forward-directed time travel (for example, biostasis) is "very likely"; and, (2) past-directed time travel (for example, time viewing) is "likely" (given reasonable assumptions explicitly stated by him).

Dr. Tandy is dedicated to encouraging scientific advancements while helping humanity to proactively prevent doomsday and to wisely foresee and manage catastrophic risks and possible misuse of increasingly powerful technologies, including bio-technology, nano-technology, and info-technology, as we move towards a technological singularity. In some situations it is both desirable and feasible to ban technology or to promote technology or to relinquish technological capacity in favor of the public interest, so as to prevent doomsday and wisely manage risks and opportunities. (For example, Dr. Tandy opposes action to build space-based weapons. For example, Dr. Tandy favors action to build self-sufficient self-replicating extraterrestrial green-habitat communities. For example, Dr. Tandy opposes the U.S. government recently purposely posting on the internet a recipe for the 1918 flu virus. For example, Dr. Tandy opposes the U.S. government recently accidentally posting on the internet a guidebook on how to build an atomic bomb.)

Dr. Charles Tandy is an Associate Professor of Humanities, and a Senior Faculty Research Fellow in Bioethics, at Fooyin University (Taiwan). He serves there on the Faculty of History and Philosophy and on the Medical Humanities Research Faculty. Dr. Tandy's websites include <www.DoctorTandy.com> and <www.SEGITs.com>. His email address is <tandy@ria.edu>.

ABOUT THE DEATH AND ANTI-DEATH SERIES

The Death And Anti-Death Series By Ria University Press (www.ria.edu/rup) discusses issues and controversies related to death, life extension, and anti-death, broadly construed. A variety of differing points of view are presented and argued. The following volumes have been published:

Death And Anti-Death, Volume 1:
One Hundred Years After N. F. Fedorov (1829-1903)

Charles Tandy, Ph.D., Editor
Volume One ISBN-13: 978-0-9743472-0-2 (Hardback)
Volume One ISBN-10: 0-9743472-0-5 (Hardback)
Published 2003
Distributed By Ingram

The anthology discusses a number of interdisciplinary cultural, psychological, metaphysical, and moral issues and controversies related to death, life extension, and anti-death. This first volume in the series is in honor of the 19th century Russian philosopher N. F. Fedorov. (Some of the contributions are about Fedorov; most are not.) Each of the 17 chapters includes a selected or short bibliography. The anthology also contains an Introduction and an Index -- as well as an Abstracts section that serves as an extended table of contents.

A variety of differing points of view are presented and argued. Most of the 400-plus pages consist of contributions unique to this volume. Although of interest to the general reader, the anthology functions well as a textbook for university courses in culture studies, death-related controversies, ethics, futuristics, humanities, interdisciplinary studies, life extension issues, metaphysics, and psychology.

The titles of the contributions are as follows:

Death And Anti-Death, Volume 2:
Two Hundred Years After Kant, Fifty Years After Turing

Charles Tandy, Ph.D., Editor
Volume Two ISBN-13: 978-0-9743472-2-6 (Hardback)
Volume Two ISBN-10: 0-9743472-2-1 (Hardback)
Published 2004
Distributed By Ingram

The following contributions are original to this volume of the Death And Anti-Death Series By Ria University Press:

- **Is The Universe Immortal?: Is Cosmic Evolution Never-Ending?** by Charles Tandy

- **Death As Metaphor** by Lawrence Kimmel

- **Fantasies Of Immortality** by Werner J. Wagner

- **What Will The Immortals Eat?** by George M. Young

- **Cultural Death Understanding** by Anthony S. Dawber

- **Death And Immortality: Plato, Aristotle, Aquinas And Descartes On The Soul** by Carol O'Brien

- **Against The Immortality Of The Soul** by Matt McCormick

- **Why Death Is (Probably) Bad For You: A Common Sense Approach** by R.C.W. Ettinger

- **Resurrecting Kant's Postulate Of Immortality** by Scott R. Stroud

- **Immortality and Finitude: Kant's Moral Argument Reconsidered** by Douglas Burnham

- **Death, Harm, And The Deprivation Theory** by Jack Li (Author now known as Jack Lee)

- **To Be Or Not To Be: The Zombie In The Computer** by R.C.W. Ettinger

- **The Future Of Human Evolution** by Nick Bostrom

- **Earthlings Get Off Your Ass Now!: Becoming Person, Learning Community** by Charles Tandy

---

Death And Anti-Death, Volume 3:
Fifty Years After Einstein,
One Hundred Fifty Years After Kierkegaard

Charles Tandy, Ph.D., Editor
Volume Three ISBN-13: 978-0-9743472-6-4 (Hardback)
Volume Three ISBN-10: 0-9743472-6-4 (Hardback)
Published 2005
Distributed By Ingram

Volume Three in the Death And Anti-Death Series By Ria University Press is in honor of Albert Einstein and Soren Kierkegaard. The chapters do not necessarily mention Einstein or Kierkegaard. The 17 chapters (by professional philosophers and other professional scholars) are directed to issues related to death, life extension, and anti-death. Most of the 400-plus pages consist of scholarship unique to this volume. Includes index.

The titles of the 17 chapters are as follows:

- **Death And Life Support Systems: A Novel Cultural Exploration** by Giorgio Baruchello

- **Recent Developments In The Ethics, Science, And Politics Of Life-Extension** by Nick Bostrom

- **Return To A Pristine Ecosphere Via Molecular Nanotechnology** by Sinclair T. Wang

- **Fedorov's Legacy: The Cosmist View Of Man's Role In The Universe** by George M. Young

---

Death And Anti-Death, Volume 4:
Twenty Years After De Beauvoir,
Thirty Years After Heidegger

Charles Tandy, Ph.D., Editor
Volume Four ISBN-13: 978-0-9743472-8-8 (Hardback)
Volume Four ISBN-10: 0-9743472-8-0 (Hardback)
Published 2006
Distributed By Ingram

Volume Four, as indicated by the anthology's subtitle, is in honor of Simone de Beauvoir (1908-1986) and Martin Heidegger (1889-1976). The chapters do not necessarily mention Simone de Beauvoir or Martin Heidegger. The 16 chapters (by professional philosophers and other professional scholars) are directed to issues related to death, life extension, and anti-death. Most of the 400-plus pages consist of scholarship unique to this volume. Includes index.

The titles of the 16 chapters are as follows:

1. Mechanism, Galileo's *Animale* And Heidegger's *Gestell*: Reflections On The Lifelessness Of Modern Science by Giorgio Baruchello

2. Simone De Beauvoir by Debra Bergoffen

16. Embryo Cloning: Current State Of The Medical Art And Its Far-Reaching Consequences For Multiple Applications by Panayiotis M. Zavos

---

Death And Anti-Death, Volume 5:
Thirty Years After Loren Eiseley (1907-1977)

Charles Tandy, Ph.D., Editor
ISBN 978-1-934297-02-5 (Hardback)
Published 2007
Distributed By Ingram

Volume 5, as indicated by the anthology's subtitle, is in honor of Loren Eiseley (1907-1977). The chapters do not necessarily mention him. The chapters (by professional philosophers and other professional scholars) are directed to issues related to death, life extension, and anti-death. Most of the contributions consist of scholarship unique to this volume. As was the case with all previous volumes in the Death And Anti-Death Series By Ria University Press, the anthology includes an Index as well as an Abstracts section that serves as an extended table of contents. (With Volume 5, you will also find a new section entitled BRIEF COMMUNICATIONS.) The 17 chapter titles are as follows:

1. Asking The Unaskable Question – Do People Have The Right NOT To Die? by Marcus Barber

2. Deadly Economics: Reflections On The Neoclassical Paradigm by Giorgio Baruchello

3. A Frozen Future? Cryonics As A Gamble by Gregory Benford

16. Teleological Causes And The Possibilities Of Personhood by Charles Tandy

17. Terrestrial Peoples, Extraterrestrial Persons by Charles Tandy

---

Death And Anti-Death, Volume 6:
Thirty Years After Kurt Gödel (1906-1978)

Charles Tandy, Ph.D., Editor
ISBN 978-1-934297-03-2 (Hardback)
Published 2008
Distributed By Ingram

Volume 6, as indicated by the anthology's subtitle, is in honor of Kurt Gödel (1906-1978). The chapters do not necessarily mention him. The chapters (by professional philosophers and other professional scholars) are directed to issues related to death, life extension, and anti-death, broadly construed. Most of the contributions consist of scholarship unique to this volume. As was the case with all previous volumes in the Death And Anti-Death Series By Ria University Press, the anthology includes an Index as well as an Abstracts section serving as an extended table of contents. (Volume 6 also includes a BRIEF COMMUNICATIONS section.) The ten chapters are entitled as follows:

1. Life And Death Economics: A Dialogue by Giorgio Baruchello and Valerio Lintner (pages 33-52)

2. Charles Hartshorne by Daniel A. Dombrowski (pages 53-78)

ABOUT THE CULTURAL CLASSICS SERIES

The Cultural Classics Series By Ria University Press (www.ria.edu/rup) seeks to reprint and keep in print books deemed to be cultural classics. Your suggestions for future volumes are welcomed. The following volumes in the series have already been published:

The Prospect of Immortality
Robert C. W. Ettinger

(Charles Tandy, Ph.D., Editor)
ISBN-13: 978-0-9743472-3-3 (Hardback)
ISBN-10: 0-9743472-3-X (Hardback)
Published 2005
Distributed By Ingram

This 2005 edition contains an exact replica copy of the complete first edition of Robert Ettinger's 1964 cultural classic, *The Prospect of Immortality*. Additional materials include three original (2005) paper contributions: (1) "Ettinger's 1964 Thesis: Indefinitely Extended And Enhanced Life (Immortality) Is Probably Already Here Via Experimental Long-Term Suspended Animation" (By Charles Tandy); (2) "The State of Cryonics -- 2005" (By Jim Yount); and, (3) "A Brief History of Cryonics" (By R. Michael Perry). (Note: James Bedford began his journey as "the first cryonaut" on January 12, 1967; as of 2005, he and many others remain in cryonic hibernation.)

According to Ettinger, cryonic hibernation (experimental long-term suspended animation) of humans may provide a "door into summer" unlike any season previously known. Such patients (individuals and families in cryonic hibernation) may yet experience the transhuman condition.

Ettinger argues for his belief in "the possibility of limitless life for our generation." We should become aware of the incorrect, distorted, and oversimplified ideas presented in the popular media about cryonics. He believes that the cool logic and scientific evidence he presents should lead us to forget the horror movies and urban legends and embrace great expectations.

Man into Superman
The Startling Potential of Human Evolution - And How to Be Part of It
R. C. W. Ettinger

(Charles Tandy, Ph.D., Editor)
ISBN-13: 978-0-9743472-4-0 (Hardback)
ISBN-10: 0-9743472-4-8 (Hardback)
Published 2005
Distributed By Ingram

In the 1960s Ettinger founded the cryonics (cryonic hibernation) movement and authored *The Prospect of Immortality*. In the 1970s Ettinger helped initiate the transhumanist revolution with his *Man into Superman*. Ettinger sees "discontinuity in history, with mortality and humanity on one side -- on the other immortality and transhumanity."

Cryonic hibernation (experimental long-term suspended animation) of humans may provide a "door into summer" unlike any season previously known. Such patients (individuals and families in cryonic hibernation) may yet experience the transhuman condition. Ettinger argues for his belief in "the possibility of limitless life for our generation." We should become aware of the incorrect, distorted, and oversimplified ideas presented in the popular media about cryonics and transhumanism. Ettinger believes that the cool logic and scientific evidence he presents should lead us to forget the horror movies and urban legends and embrace great expectations.

This 2005 edition contains an exact replica copy of the complete first edition of Ettinger's 1972 cultural classic, *Man into Superman*. Additional materials (three articles) include comments by others -- "Developments In Transhumanism 1972-2005" -- written especially for this 21st century edition. For example, Dr. Nick Bostrom, a professional philosopher at the University of Oxford (UK) and a founder of the World Transhumanist Association, provides us with "A Short History of Transhumanist Thought."

Ria University Press
(www.ria.edu/rup)

Ria University Press
(www.ria.edu/rup)

Ria University Press
(www.ria.edu/rup)